AQA GCSE (9–1)
Biology
Grade 5 Booster Workbook

For Combined Science: Trilogy

Heidi Foxford

Shaista Shirazi

William Collins' dream of knowledge for all began with the publication of his first book in 1819. A self-educated mill worker, he not only enriched millions of lives, but also founded a flourishing publishing house. Today, staying true to this spirit, Collins books are packed with inspiration, innovation and practical expertise. They place you at the centre of a world of possibility and give you exactly what you need to explore it.

Collins. Freedom to teach

HarperCollins Publishers
The News Building
1 London Bridge Street
London SE1 9GF

HarperCollins *Publishers*
Macken House, 39/40 Mayor Street Upper,
Dublin 1, D01 C9W8, Ireland

Browse the complete Collins catalogue at
www.collins.co.uk

10 9 8 7 6

© HarperCollins Publishers 2018

ISBN 978-0-00-829653-7

Collins® is a registered trademark of HarperCollins Publishers Limited

www.collins.co.uk

A catalogue record for this book is available from the British Library

Commissioned by Joanna Ramsay and Rachael Harrison
Development edited by Gillian Lindsey
Project managed by Sarah Thomas, Siobhan Brown and Mike Appleton
Copy edited by Rebecca Ramsden and Aidan Gill
Proofread by Helen Bleck
Technical review by Rich Cutler
Typeset by Jouve India Pvt Ltd.,
Cover design by We are Laura and Jouve
Cover image: Arsenis Spyros/Shutterstock, Remoau/Shutterstock
Production by Tina Paul

Printed and Bound in the UK by Ashord Colour Press Ltd

This book is produced from independently certified FSC™ pa
to ensure responsible forest management.

For more information visit: www.harpercollins.co.uk/green

HarperCollins does not warrant that www.collins.co.uk or any other website mentioned in this title will be provided uninterrupted, that any website will be error free, that defects will be corrected, or that the website or the server that makes it available are free of viruses or bugs. For full terms and conditions please refer to the site terms provided on the website.

Contents

Introduction

This workbook will help you build your confidence in answering Biology questions for GCSE Combined Science.

It gives you practice in using key scientific words, writing longer answers, answering synoptic questions as well as applying knowledge and analysing information.

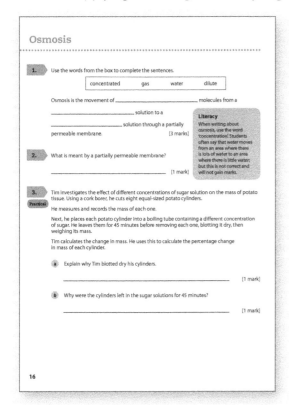

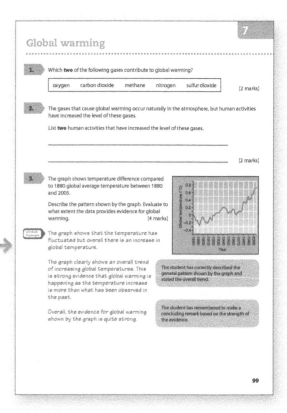

Learn how to answer test questions with annotated worked examples.

This will help you develop the skills you need to answer questions.

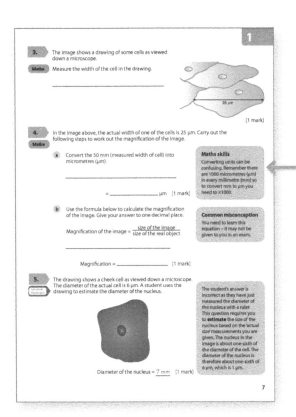

3. The image shows a drawing of some cells as viewed down a microscope.

Maths | Measure the width of the cell in the drawing.

_____

25 μm

[1 mark]

4. In the image above, the actual width of one of the cells is 25 μm. Carry out the following steps to work out the magnification of the image.

Maths

  a Convert the 50 mm (measured width of cell) into micrometres (μm).

_____

= _____ μm   [1 mark]

Maths skills
Converting units can be confusing. Remember there are 1000 micrometres (μm) in every millimetre (mm) so to convert mm to μm you need to ×1000.

  b Use the formula below to calculate the magnification of the image. Give your answer to one decimal place.

Magnification of the image = size of the image / size of the real object

_____

Magnification = _____   [1 mark]

Common misconception
You need to learn this equation – it may not be given to you in an exam.

5. The drawing shows a cheek cell as viewed down a microscope. The diameter of the actual cell is 6 μm. A student uses the drawing to estimate the diameter of the nucleus.

Diameter of the nucleus = 7 mm   [1 mark]

The student's answer is incorrect as they have just measured the diameter of the nucleus with a ruler. This question requires you to **estimate** the size of the nucleus based on the 'actual size' measurements you are given. The nucleus in the image is about one-sixth of the diameter of the cell. The diameter of the nucleus is therefore about one-sixth of 6 μm, which is 1 μm.

7

There are lots of hints and tips to help you out. Look out for tips on how to decode command words, key tips for required practicals and maths skills, and common misconceptions.

The amount of support gradually decreases throughout the workbook. As you build your skills you should be able to complete more of the questions yourself.

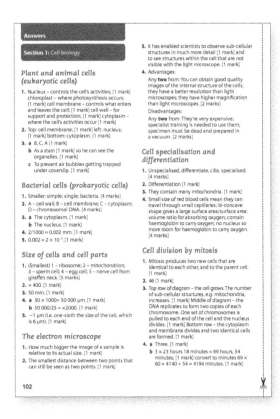

Answers

Section 1: Cell biology

Plant and animal cells (eukaryotic cells)

1. Nucleus – controls the cell's activities; [1 mark] chloroplast – where photosynthesis occurs; [1 mark] cell membrane – controls what enters and leaves the cell; [1 mark] cell wall – for support and protection; [1 mark] cytoplasm – where the cell's activities occur [1 mark]
2. Top: cell membrane; [1 mark] left: nucleus; [1 mark] bottom: cytoplasm. [1 mark]
3. **a** B, C, A [1 mark]
   b As a stain [1 mark] so he can see the organelles. [1 mark]
   c To prevent air bubbles getting trapped under coverslip. [1 mark]

Bacterial cells (prokaryotic cells)

1. Smaller; simple; single; bacteria. [4 marks]
2. A – cell wall; B – cell membrane; C – cytoplasm; D – chromosomal DNA. [4 marks]
3. **a** The cytoplasm. [1 mark]
   b The nucleus. [1 mark]
4. 2/1000 = 0.002 mm. [1 mark]
5. 0.002 = 2 × 10^{-3}. [1 mark]

Size of cells and cell parts

1. (Smallest) 1 – ribosome; 2 – mitochondrion; 3 – sperm cell; 4 – egg cell; 5 – nerve cell from giraffe's neck. [5 marks]
2. × 400. [1 mark]
3. 50 mm. [1 mark]
4. **a** 50 × 1000= 50 000 μm. [1 mark]
   b 50 000/25 = ×2000. [1 mark]
5. ~1 μm (i.e. one-sixth the size of the cell, which is 6 μm). [1 mark]

The electron microscope

1. How much bigger the image of a sample is relative to its actual size. [1 mark]
2. The smallest distance between two points that can still be seen as two points. [1 mark]

3. It has enabled scientists to observe sub-cellular structures in much more detail [1 mark] and to see structures within the cell that are not visible with the light microscope. [1 mark]
4. Advantages:
   Any **two** from: You can obtain good quality images of the internal structure of the cells; they have a better resolution than light microscopes; they have higher magnification than light microscopes. [2 marks]
   Disadvantages:
   Any **two** from: They're very expensive; specialist training is needed to use them; specimen must be dead and prepared in a vacuum. [2 marks]

Cell specialisation and differentiation

1. Unspecialised, differentiate, cilia, specialised. [4 marks]
2. Differentiation [1 mark]
3. They contain many mitochondria. [1 mark]
4. Small size of red blood cells mean they can travel through small capillaries; bi-concave shape gives a large surface area/surface area: volume ratio for absorbing oxygen; contain haemoglobin to carry oxygen; no nucleus so more room for haemoglobin to carry oxygen. [4 marks]

Cell division by mitosis

1. Mitosis produces two new cells that are identical to each other, and to the parent cell. [1 mark]
2. 46 [1 mark]
3. Top row of diagram – the cell grows. The number of sub-cellular structures, e.g. mitochondria, increases. [1 mark] Middle of diagram – the DNA replicates to form two copies of each chromosome. One set of chromosomes is pulled to each end of the cell and the nucleus divides. [1 mark] Bottom row – the cytoplasm and membrane divides and two identical cells are formed. [1 mark]
4. **a** Three. [1 mark]
   b 3 × 23 hours 18 minutes = 69 hours, 54 minutes; [1 mark] convert to minutes 69 × 60 = 4140 + 54 = 4194 minutes. [1 mark]

102

There are answers to all the questions at the back of the book. You can check your answers yourself or your teacher might tear them out and give them to you later to mark your work.

Plant and animal cells (eukaryotic cells)

1. ▷ Draw **one** line from each part of a cell to its correct function.

Nucleus	Where photosynthesis occurs
Chloroplast	Controls what enters and leaves the cell
Cell membrane	Where the cell's activities occur
Cell wall	Controls the cell's activities
Cytoplasm	For support and protection

[5 marks]

2. ▷ The diagram shows an animal cell. Complete the labels on the diagram.

[3 marks]

3. ▷ Freddie wants to look at an onion cell under a light microscope. He places a thin
layer of onion skin on a slide.

a Write the letters **A–C** in the correct order to show how he should prepare the slide.

A: Use a paper towel to absorb any liquid that spreads out from under the coverslip.

B: Use a pipette to add a small drop of iodine solution onto the slide.

C: Place a coverslip onto the slide by lowering it carefully from one side.

_____ [1 mark]

b Why does Freddie use iodine solution?

_____ [2 marks]

c State why it is important to 'lower the coverslip carefully from one side'.

_____ [1 mark]

Bacterial cells (prokaryotic cells)

1. Use the words from the box to complete the sentences.

| multi | single | smaller | larger | plants | bacteria | simple | complex |

Prokaryotic cells are _____ and

more _____ than eukaryotic

cells. They are _____

-celled organisms. Prokaryotes include

_____ and archaea.

Literacy

Most scientific words have a Greek or Latin origin which can help you remember the meaning. The prefix 'pro' means 'before' as prokaryotes are the earliest form of life on Earth.

[4 marks]

2. The diagram shows a bacterial cell. Name the parts A, B, C and D.

A: _____

B: _____

C: _____

D: _____

[4 marks]

3. Name the part of the cell where genetic material is found, in

1 prokaryotic cells; _____ [1 mark]

2 eukaryotic cells. _____ [1 mark]

4. *Escherichia coli* is a rod-shaped bacterium. Each bacterium measures approximately 2 μm in length. Convert this figure into mm. Show your workings.

Maths

_____

= _____ mm   [1 mark]

5. Give your answer to question ④ in standard form.

Maths

_____

= _____ mm   [1 mark]

Size of cells and cell parts

1. Owen does some research and finds out the sizes of different cells and organelles. Put the cells and organelles in order of size, from smallest (1) to biggest (5).

Maths

Cells/organelle type	Size	Order (1–5)
Egg cell	0.12 mm	
Sperm cell	40.0 μm	
Ribosome	20 nm	
Nerve cell from giraffe's neck	3 m	
Mitochondrion	2 μm	

[5 marks]

2. Baljit is using a microscope. The magnification of the eyepiece lens is ×10 and the magnification of the objective lens is ×40. What is the total magnification?
Tick **one** box.

Maths

☐ ×40   ☐ ×50   ☐ ×400   ☐ ×4000

[1 mark]

3. The image shows a drawing of some cells as viewed down a microscope.

Maths Measure the width of the cell in the drawing.

_____

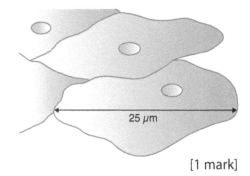

25 μm

[1 mark]

4. In the image above, the actual width of one of the cells is 25 μm. Carry out the following steps to work out the magnification of the image.

Maths

a Convert the 50 mm (measured width of cell) into micrometres (μm).

_____

= _____ μm   [1 mark]

Maths skills

Converting units can be confusing. Remember there are 1000 micrometres (μm) in every millimetre (mm) so to convert mm to μm you need to ×1000.

b Use the formula below to calculate the magnification of the image. Give your answer to one decimal place.

$$\text{Magnification of the image} = \frac{\text{size of the image}}{\text{size of the real object}}$$

_____

Magnification = _____ [1 mark]

Common misconception

You need to learn this equation – it may not be given to you in an exam.

5. The drawing shows a cheek cell as viewed down a microscope. The diameter of the actual cell is 6 μm. A student uses the drawing to estimate the diameter of the nucleus.

Worked Example

Diameter of the nucleus = 7 mm   [1 mark]

The student's answer is incorrect as they have just measured the diameter of the nucleus with a ruler. This question requires you to **estimate** the size of the nucleus based on the 'actual size' measurements you are given. The nucleus in the image is about one-sixth of the diameter of the cell. The diameter of the nucleus is therefore about one-sixth of 6 μm, which is 1 μm.

The electron microscope

1. Describe what is meant by magnification.

_____ [1 mark]

2. Which of the following is the **best** definition for 'resolution'?
Tick **one** box.

☐ The amount of colour seen in an image.

☐ The smallest distance between two points that can still be seen as two points.

☐ The smallest object that can be observed using a microscope.

☐ The amount by which a microscope can magnify. [1 mark]

3. Explain how electron microscopy has increased our understanding of
sub-cellular structures.

_____

_____ [2 marks]

4. Research laboratories often use electron microscopes rather than the light
microscopes used in schools. Describe the advantages and disadvantages
of using electron microscopes.

_____

_____

_____

_____ [4 marks]

Cell specialisation and differentiation

1. Use words from the box to complete the sentences.

| specialised | unspecialised | differentiate | mutated | cilia |

We start our lives as a single fertilised egg which grows to become an embryo. At this

point the cells are _____, but have the potential to

_____ into any of 200 or so specialised cell types

that make up the human body. As a cell differentiates it develops sub-cellular

structures such as _____ to enable it to carry

out a certain function. It has become a _____ cell. [4 marks]

2. The process of forming specialised cells to make them
suitable for their function is called (tick **one** box):

◻ Adaptation   ◻ Differentiation

◻ Specialisation   ◻ Mitosis   [1 mark]

> **Literacy**
> Make sure you can explain the difference between key words that are closely linked. Cell differentiation is the process of how a specialised cell is made, whereas cell specialisation is how certain cells are adapted to a function.

3. How are muscle cells adapted to release a lot of energy?

_____

_____ [1 mark]

4. The diagram shows a red blood cell.

Describe and explain how red blood cells are adapted for
the efficient uptake and transport of oxygen.

_____

_____

_____

_____

cytoplasm containing
haemoglobin, which
transports oxygen

cell membrane

[4 marks]

Cell division by mitosis

1. Which of the following statements **best** describes mitosis? Tick **one** box.

☐ Mitosis produces two new cells that are identical to each other, and to the parent cell.

☐ Mitosis produces one old cell and one new cell that are identical.

☐ Mitosis produces two new cells that are genetically different.

☐ Mitosis produces some cells that have double the amount of chromosomes.   [1 mark]

2. Give the total number of chromosomes you would expect to find in a human cell produced by mitosis.

_____   [1 mark]

3. Draw **one** line from each stage of the cell cycle shown in the diagram to link it with the correct description of the cell cycle in the table.

| The cytoplasm and membrane divide and two identical cells are formed. |

| The DNA replicates to form two copies of each chromosome. One set of chromosomes is pulled to each end of the cell and the nucleus divides. |

| The cell grows. The number of sub-cellular structures, e.g. mitochondria, increases. |

4. A human cell completed the cell cycle in 23 hours 18 minutes.

Maths

a How many cell cycles would be needed to produce eight cells?

_____

Number of cell cycles = _____   [1 mark]

Maths skills

Questions that require you to carry out calculations may ask you to convert units, e.g. from hours to minutes or vice versa. Make sure you read the questions carefully. Underline the units you are asked to answer the question in.

b Calculate the time required in minutes for this cell to produce eight cells.

_____

Time required = _____ minutes   [2 marks]

5. In the cell cycle, where are chromosomes found before mitosis starts?

_____   [1 mark]

6. State **two** things that must happen during the cell cycle before mitosis can begin.

_____   [1 mark]

Stem cells

1. Use the words from the box below to complete the sentences.

embryos children differentiate mutate bone-marrow muscle

Most cells in your body are differentiated for one particular function, but some are completely unspecialised. These are called stem cells. They can

_____ into many different types of cells when

they are needed. Human stem cells are found in _____

and in some adult tissue such as _____.   [3 marks]

2. A human stem cell can develop into what? Tick **one** box.

☐ Gametes   ☐ Some cells

☐ Only nerve cells   ☐ Any type of human cell   [1 mark]

3. Plants also produce stem cells. What is the name of the tissue that produces stem cells in plants?

_____   [1 mark]

4. One way that scientists can investigate stem cells is to use spare embryos from *in-vitro* fertilisation (IVF) treatment. A fertilised egg develops into an embryo and stem cells are removed for research. Give **one** reason why people might object to research on embryonic stem cells.

_____ [1 mark]

5. To avoid ethical issues concerning embryos, stem cells in the future may be taken from which source? Tick **one** box.

☐ Blood   ☐ Umbilical cord

☐ Rats   ☐ Lungs of an adult   [1 mark]

6. Stem cell transplants could help people with paralysis. What type of cell would need to be grown from stem cells to help a person with paralysis?

_____ [1 mark]

7. Evaluate the risks and benefits associated with the use of stem cells for medical treatments.

Worked Example

A student's answer to this question is shown below.

I think we should be using stem cells for treatments because it can make people's illness or conditions better and can improve their quality of life. Stem cells can be used to treat and reduce suffering for many conditions like Parkinson's and diabetes or after treatment for cancer. Another benefit is that there is hardly any risk of rejection.

Overall, if a person is ill and can have stem cell treatment, it is better to take the risk because the benefits of a better life would always outweigh the risks.

The student has correctly outlined the benefits which would have gained 2 marks, but they have not outlined the risks (unknown long term side effects, chance of rejection if stem cells are not from same person). They have lost marks by not writing about what the risks are.

The student has evaluated the information they have presented and given an overall opinion based on this, so would get a mark for evaluating.

[6 marks]

Diffusion

1. Which of the following statements **best** describes the process of diffusion? Tick **one** box.

Diffusion is the movement of:

☐ molecules from an area of high concentration to an area of low concentration.

☐ molecules from an area of low concentration to an area of high concentration.

☐ gas and water molecules that move randomly.

☐ water molecules that move against a concentration gradient. [1 mark]

2. The diagram shows a cell surrounded by oxygen. Draw an arrow on the diagram to show which direction the oxygen particles will move in.

cell membrane, which
is permeable to oxygen

high concentration
of oxygen

low concentration
of oxygen

[1 mark]

3. Name **two** factors that affect the rate of diffusion.

_____

_____ [2 marks]

4. Red blood cells are **not** normally able to diffuse from the blood into the surrounding body tissues, but substances such as oxygen or glucose are. Explain why this is.

_____ [1 mark]

5. Explain why the rate of diffusion of carbon dioxide into stomata on a leaf is higher on a warm day.

_____

_____ [2 marks]

6. Describe and explain the role of diffusion in gas exchange in the lungs.

_____

_____

_____

_____

_____ [6 marks]

Exchange surfaces in animals

1. The surface area of a cell affects the rate at which particles can enter and leave the cell. The table shows different-sized cubes that represent cells. Complete the table by calculating the surface area, volume and surface area to volume ratio (SA:V) for the 3 × 3 × 3 cm cube.

Cubes representing cells (cm)	Surface area (cm²)	Volume (cm³)	SA:V
1 cm cube	6	1	6:1
2 cm cube	24	8	3:1
3 cm cube	_____	_____	_____

Maths
You need to know how to calculate the surface area to volume ratio. Make sure you learn the formula:

Surface area to volume ratio $= \dfrac{\text{surface area}}{\text{volume}}$

_____ [3 marks]

2. The diagram shows **two** different single-celled organisms.

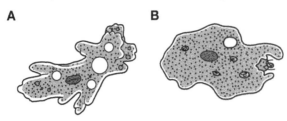

A   B

14

Explain which cell, **A** or **B**, could support a faster rate of diffusion of nutrients.

_____

_____ [2 marks]

3. Single-celled organisms do **not** have lungs, but large multicellular organisms like humans do have lungs. Explain why.

_____

_____

_____

_____ [4 marks]

4. Using the diagram, describe how the surface of the small intestine is adapted for exchanging materials. Explain how it is adapted for this.

The villi

blood vesels

villi

small intestine wall

_____

_____

_____

_____

_____ [3 marks]

Osmosis

1. Use the words from the box to complete the sentences.

concentrated	gas	water	dilute

Osmosis is the movement of _____ molecules from a

_____ solution to a

_____ solution through a partially

permeable membrane. **[3 marks]**

> **Literacy**
> When writing about osmosis, use the word 'concentration'. Students often say that water moves from an area where there is lots of water to an area where there is little water; but this is not correct and will not gain marks.

2. What is meant by a partially permeable membrane?

_____ **[1 mark]**

3. Tim investigates the effect of different concentrations of sugar solution on the mass of potato tissue. Using a cork borer, he cuts eight equal-sized potato cylinders.

Practical

He measures and records the mass of each one.

Next, he places each potato cylinder into a boiling tube containing a different concentration of sugar. He leaves them for 45 minutes before removing each one, blotting it dry, then weighing its mass.

Tim calculates the change in mass. He uses this to calculate the percentage change in mass of each cylinder.

a Explain why Tim blotted dry his cylinders.

_____ **[1 mark]**

b Why were the cylinders left in the sugar solutions for 45 minutes?

_____ **[1 mark]**

Tim's results are shown below.

Boiling tube number	1	2	3	4	5	6	7	8
Concentration of sugar solution (M)	0.1	0.2	0.3	0.4	0.5	0.6	0.7	0.8
% change in mass of potato cylinder	6.5	5.0	3.0	0.0	−3.5	−5	−7.5	−9.5

Maths **c** Draw a graph to show concentration of sugar solution against percentage change in mass. Add a line of best fit.

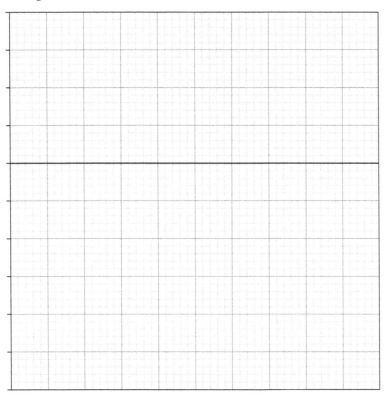

[3 marks]

Maths **d** Use your graph to estimate the percentage change of mass if a 0.9 M concentration of sugar solution is used.

Percentage change = _____ % [1 mark]

e Tim concludes that the concentration of the sugar solution inside potato cells is about 0.4 M. Do you agree? Explain your answer.

_____

_____ [2 marks]

Active transport

1. Use the words from the box to complete the sentences.

heat	higher	against	along	energy	lower

Active transport is the process by which dissolved molecules move across a cell

membrane from a _____ to a _____

concentration. In active transport, particles move _____
the concentration gradient and therefore require an input of

_____ from the cell. [4 marks]

2. Active transport allows mineral ions to be absorbed from the soil into the plant root hair cells.

Explain why active transport is necessary for ions to be transported into the root hair cells.

_____

_____ [2 marks]

3. A student is asked to compare active transport and diffusion. Their answer is shown
below. [4 marks]

Worked Example

Active transport works against a
concentration gradient whereas
diffusion works along a concentration
gradient.

Active transport requires energy from
respiration whereas diffusion does
not require energy. Both are forms of
transport that move substances.

The command word 'compare' means to
consider differences and similarities and the
student has correctly included a similarity to
gain full marks.

The student has correctly compared active
transport and diffusion. Their answer
contains the right amount of detail and good
use of scientific language.

Digestive system

1. The diagram shows the human digestive system. Complete the labels using the words below.

| large intestine | oesophagus | small intestine |

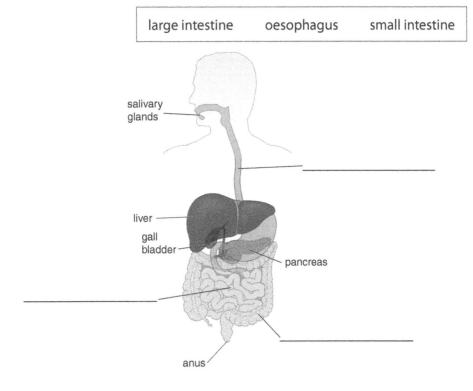

salivary glands

liver

gall bladder

pancreas

anus

[3 marks]

2. Which of the following statements **best** describes the digestive system? Tick **one** box.

☐ An organ system in which several organs work together to break down and absorb food.

☐ An organ consisting of the stomach which digests and absorbs all food.

☐ A tissue that is specialised for absorbing glucose.

☐ All of the above. [1 mark]

3. Why is digestion necessary?

_____ [1 mark]

4. Explain how physical digestion is different from chemical digestion.

_____

_____ [2 marks]

19

Digestive enzymes

1. Use words from the box to complete the sentences.

| stomach | respiration | bloodstream | large | small | bile salts | proteins |

Digestive enzymes break down food into _____ molecules that can be

absorbed into the _____ . The products of digestion are used to build

new carbohydrates, lipids and _____ . Some of the glucose is used in

_____ .   [4 marks]

2. Complete the table.

Enzyme	Site of production	Reaction
Amylase	_____	Starch to _____
_____	Stomach, pancreas, small intestine	Protein to amino acids
Lipase	Pancreas, small intestine	_____ to fatty acids and glycerol

[4 marks]

3. Draw **one** line from each test to the biological molecules that it identifies.

Benedict's test		Carbohydrates – starch
Biuret test		Carbohydrates – sugars
Iodine test		Proteins
		Lipids

[3 marks]

4. **a** Joanne conducts three different food tests on peanuts to find out what they contain. First she grinds the nuts with a pestle and mortar before adding distilled water.

Practical

Explain why she does this.

_____

_____ [2 marks]

Practical
You could be tested on any of the food tests in the exam. Make sure you know the names of the reagents, how to conduct the test, and the colour change to expect.

b Joanne wants to find out if milk contains protein. Describe a test to find out if protein is present.

_____

_____ [2 marks]

Factors affecting enzymes

1. Sarah investigates the effect of temperature on the activity of the enzyme amylase. She uses a continuous sampling technique to determine the time taken to completely digest a starch solution at a range of temperatures. Every 30 seconds she uses iodine solution to test for

Practical starch.

a Describe how Sarah will know when the starch has been broken down.

_____ [1 mark]

b Sarah uses a water bath to heat the amylase. Give **one** advantage of using a water bath rather than a Bunsen burner and beaker of water to heat the amylase.

_____ [1 mark]

c What is meant by a 'continuous sampling technique'?

_____ [1 mark]

d Sarah's results are shown in the graph.

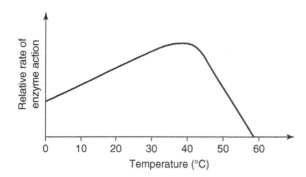

Label the graph to show the optimum temperature for amylase. [1 mark]

e Describe and explain what is happening at this point.

_____

_____

_____ [2 marks]

Heart and blood vessels

1. The diagram shows a heart.

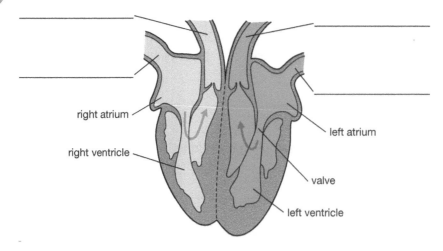

a Add labels to the diagram for the following vessels: aorta, vena cava, pulmonary artery, pulmonary vein. [4 marks]

b Add an arrow to the diagram to show where oxygenated blood from the lungs enters the heart. [1 mark]

c Add an arrow to the diagram to show where oxygenated blood leaves the heart to be pumped around the body. [1 mark]

2. Which of the following statements about pacemakers are true? Tick **two** boxes.

☐ A pacemaker is a group of cells in the right atrium of the heart.

☐ A pacemaker is located in the brain and controls the heart rate.

☐ Artificial pacemakers are used to treat faulty valves in the heart.

☐ Artificial pacemakers can be used to correct irregularities in the heart rate. [2 marks]

3. Give **two** ways in which capillaries are structurally adapted to deliver the maximum amount of oxygen and glucose to respiring cells.

_____

_____ [2 marks]

4. Compare the structure and function of an artery with that of a vein. A student's answer to this question is given below. [6 marks]

Worked Example

The arteries have thicker walls and a smaller hole inside than veins.

Veins have valves.

The arteries carry oxygenated blood away from the heart, whereas the veins carry deoxygenated blood towards the heart (except the pulmonary artery).

The arteries have pressure and a pulse but the veins have none.

Use scientific terminology: 'a smaller lumen' rather than 'a smaller hole' as these are more likely to be on the mark scheme.

Make sure your answers are clear. The word 'none' used at the end of the student answer refers to the pressure and the pulse. It would be better to write 'the veins carry blood under lower pressure and have no pulse'.

The command word 'compare' means you should consider the differences and the similarities. This answer does not contain any similarities (e.g. they both carry blood around the body) so it would not get in the 5–6-mark band.

Blood

1. Complete the table to describe the components of blood and their functions.

Component of blood	Function
Plasma	Transports substances such as hormones, antibodies, glucose, amino acids and waste substances.
Red blood cells	_____.
White blood cells	_____.
_____	Help the clotting process at wound sites.

[3 marks]

2. List **three** substances carried in the plasma.

_____ [3 marks]

3. A blood test shows that a patient has a very high white blood cell count. Suggest a reason for this.

_____ [1 mark]

4. Explain how a biconcave shape helps a red blood cell carry out its function.

_____

_____ [2 marks]

5. **a** Approximately 55% of the blood is plasma. If a person has 6 500 cm³ of blood in their body, how much would be plasma?

Maths

_____

Plasma = _____ cm³   [1 mark]

Maths **b** In 1 mm³ of blood there are about 5 000 000 red blood cells. Write this number in standard form.

_____

= _____ red blood cells   [1 mark]

Heart–lungs system

· ·

1. Which of the following **best** describes the passage of air into the lungs? Tick **one** box.

☐ Bronchus, bronchioles, trachea, alveoli.

☐ Bronchus, bronchioles, alveoli, trachea.

☐ Trachea, bronchiole, bronchus, alveoli.

☐ Trachea, bronchus, bronchiole, alveoli.   [1 mark]

2. Why is a human described as having a double circulatory system?

_____

_____   [2 marks]

3. The diagram shows gas exchange in the alveoli.

Synoptic

bronchiole

carbon dioxide
oxygen diffuses
diffuses out
in

alveolus
(plural: alveoli)

red blood cell   blood capillary

Describe how the alveoli are adapted for efficient diffusion of oxygen and carbon dioxide. Explain how these adaptations help with diffusion.

_____

_____

Common misconception

Breathing is ventilation, or the movement of air in and out of the lungs. Respiration is **not** the same as breathing. Respiration is the release of energy from glucose, and it takes place in cells.

_____

_____ [4 marks]

Coronary heart disease

....

1. Why is coronary heart disease described as a non-communicable disease?

_____ [1 mark]

2. Write the letters in the correct order to describe the order of events in coronary heart disease.

A	The coronary arteries become narrow.
B	Heart muscle cells become so starved of oxygen that they stop contracting.
C	Layers of fatty material build up inside the coronary arteries.
D	Less oxygen gets to the heart muscle cells around the affected coronary artery.

_____ [4 marks]

3. Faulty heart valves do not open properly or they leak. Explain what effect this might have on the heart.

_____

_____ [2 marks]

4. Artificial valves may be used to replace faulty valves. Give **one** advantage and **one** disadvantage of using an artificial valve to treat a patient with a faulty valve.

Advantage: _____

Disadvantage: _____ [2 marks]

5. Barbara has coronary heart disease. She will have an operation to put a stent in her coronary artery. She will also take statins to control her blood cholesterol levels. Explain how these treatments may prevent a heart attack.

Stent: _____

Statins: _____ [2 marks]

6. Lifestyle changes are recommended for patients having stents, replacement valves or heart transplants. Suggest **two** pieces of lifestyle advice a doctor might give to a patient receiving treatment for cardiovascular disease.

_____

_____ [2 marks]

7. A heart transplant is an operation to replace a damaged heart with a healthy heart from a donor who has recently died. A student was asked to evaluate the benefits and the risks associated with this form of treatment. [5 marks]

Worked Example

The benefits are that with a heart transplant the person will live longer. Also they will have a better quality of life because they will have more energy and strength with a new heart.

> The student has correctly outlined the benefits of a heart transplant.

The risks are that the person has to have surgery which is dangerous if it goes wrong. The person could bleed to death or get an infection from the operation. Also, the person would also have to take anti-rejection drugs which might have side effects.

> The risks have been correctly identified.

Overall, the benefits outweigh the risks because the person will be alive longer with a heart transplant. The person would probably die sooner if they did not have the transplant.

> This answer would gain full marks. The student has understood the question and the command word 'evaluate', which requires weighing up the information to come to a judgement.

Risk factors for non-infectious diseases

1. Which of the following are the biggest risk factors for non-communicable diseases? Tick **one** box.

☐ Exposure to air pollution, unhealthy diet and the harmful use of alcohol.

☐ Tobacco use, malnutrition and exposure to carcinogens.

☐ Tobacco use, physical inactivity, unhealthy diet and the harmful use of alcohol.

☐ Exposure to radiation, harmful intake of caffeine and lack of sleep. [1 mark]

2. Cardiovascular disease is one of the biggest causes of premature death in the UK. Name **three** risk factors for cardiovascular disease.

1 _____

2 _____

3 _____ [3 marks]

3. Suggest why is it difficult to prove that a non-communicable disease is caused by **one** particular factor.

_____

_____ [2 marks]

4.

Maths

The graph shows the number of deaths caused by coronary heart disease for men and women in different age categories. Use the information in the graph to answer the following questions.

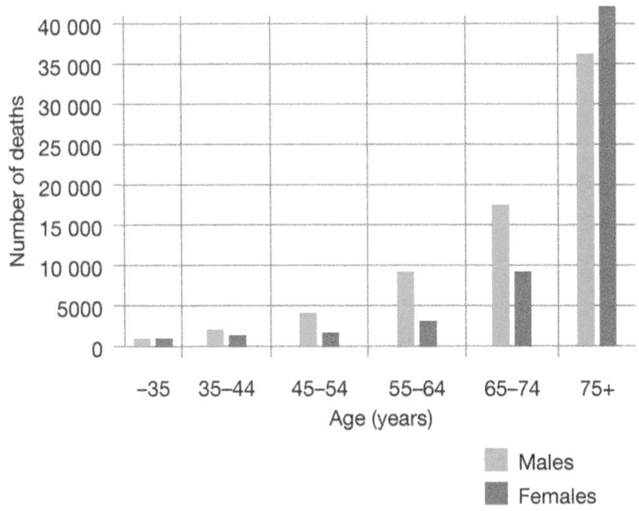

a Approximately how many deaths caused by coronary heart disease occur in the 55–64 age group for females? _____ [1 mark]

b Which of the following statements could be concluded from the data? Tick **one** box.

☐ More males suffer heart attacks than females in every age group.

☐ More older people suffer coronary heart disease than younger people.

☐ Children do not suffer from coronary heart disease.

☐ Men eat more fatty foods than women. [1 mark]

c In most age groups more men than women die of coronary heart disease. Suggest a reason why there are more coronary heart disease deaths in **women aged 75 and above** than for men.

_____ [1 mark]

Cancer

1. Use the words from the box to complete the sentences.

| limited | mutation | gland | tumour | catalyst | lifestyle | strict | brain |

Body cells normally divide under _____ control. Sometimes a

_____ causes this control to be lost. When cells divide too often they form a

_____ . This can be a result of things in our _____ or genetic
risk factors. [4 marks]

2. Give **three** risk factors associated with cancer.

_____

_____ [3 marks]

3. Which of the following statements about cancer are true? Tick **two** boxes.

☐ Tumours always start in the brain and then spread to other parts of the body.

☐ Viruses living in cells can be the trigger for certain cancers.

☐ It only takes one mutation in the DNA to trigger cancer.

☐ As a tumour grows, cancer cells can detach and spread to other parts of the body.

☐ All tumours are cancerous. [2 marks]

4. What is a carcinogen?

_____ [1 mark]

5. Describe **two** differences between a benign and a malignant tumour.

_____

_____

_____ [2 marks]

Leaves as plant organs

1. The diagram shows the internal structure of a leaf. Use words from the box to label the diagram.

| palisade cell | spongy mesophyll cell | stoma | guard cell | chlorophyll |
| vacuole | xylem | | meristem tissue | |

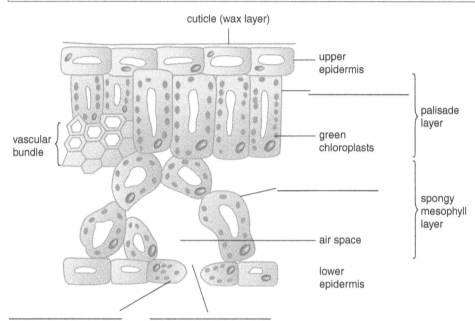

[4 marks]

2. Why is the leaf described as a plant 'organ'?

_____ [1 mark]

Remember
You will be expected to know the difference between a cell, tissue, organ and organ system. Make sure you are clear on how they are different and how they are linked.

3. Name **two** structures found in the vascular bundles.

_____ [2 marks]

4. Which statement **best** describes where meristem tissue is found? Tick **one** box. [1 mark]

☐ The roots and shoots of a plant.   ☐ The roots of a plant.

☐ The shoots of a plant.   ☐ The xylem and phloem.

5. Leaves are adapted for capturing light for photosynthesis. Explain how the following structures are adapted to collect light.

Broad leaves: _____

Palisade cells: _____

Thin and transparent upper epidermis: _____ [3 marks]

Transpiration

1. Circle **three** factors in the box that affect the rate of transpiration.

light intensity   carbon dioxide concentration   length of roots   temperature   wind

[3 marks]

2. Which of the following statements **best** describes transpiration? Tick **one** box.

☐ The evaporation of water from the leaf.

☐ The absorption of water from the soil into the roots.

☐ The movement of dissolved sugars around the plant.

☐ The movement of water through the plant and leaves. [1 mark]

3.

Maths

Ben investigates the rate of water loss from a plant shoot using a potometer. He sets up the equipment as shown.

Ben measures the distance a bubble moves along the capillary tube to find out how much water has been lost. He finds that 9 mm³ was lost in 5 minutes. Calculate the rate of water loss from the plant in mm³/s. Show your workings.

_____

Rate of water loss = _____ mm³/s

Diagram labels: leafy shoot, reservoir, tap, rubber tubing, centimetre scale, water meniscus, capillary tube, scale 0 1 2 3 4 5 6 7 8 9 10

[2 marks]

4. Rebecca compares the transpiration rates of a plant with broad flat leaves and a plant with needles. She weighs the mass of the plants at the beginning of the experiment and then 24 hours later.

a Each plant has the same number of leaves. The plants are not watered during the 24 hours.

Give **two** other experimental conditions she should keep the same.

_____ [2 marks]

The table shows Rebecca's results.

	Plant A (broad, flat leaves)	Plant B (needles)
Mass at beginning (g)	252	137
Mass after 24 hours (g)	239	129

b Write a conclusion and explanation of Rebecca's results.

Use data/calculations in your answer.

_____

_____

_____ [3 marks]

Translocation

1. Describe what is meant by 'translocation'.

_____ [1 mark]

2. What happens to the dissolved sugars that are transported around a plant? Tick **one** box.

☐ They all get used in respiration.

☐ Some is stored and some is used for respiration.

☐ It is all used for photosynthesis to make oxygen.

☐ They are all used to make fruit. [1 mark]

3. How are root hair cells adapted for the efficient uptake of water?

_____ [1 mark]

4. Compare the structure and function of xylem and phloem.

_____

_____

_____

_____

_____

_____ [6 marks]

Literacy

You will gain more marks on a 6-mark question if the points you make are linked and the answer has a clear and logical structure. If you have time, jot down some bullet points or ideas down before you start to write. By doing this you can organise your points in a logical order and increase your chances of getting more marks.

Microorganisms and disease

1. Give **one** example of each type of disease listed below.

Non-communicable: _____

Communicable: _____ [2 marks]

2. Draw **one** line from each pathogen to the disease it causes.

Viruses		Malaria
Bacteria		Flu
Protists		Athlete's foot
Fungi		Food poisoning

[4 marks]

3. Cervical cancer occurs in the neck of the uterus.

Scientists investigated the link between cervical cancer and infection with some types of Human Papilloma Virus (HPV).

The graph shows the frequency of five different types of HPV in women who had cervical cancer.

A newspaper published an article about cervical cancer with the headline 'HPV causes cervical cancer'.

Do the data shown in the graph support this claim? Explain your answer.

_____

_____

_____

_____ [4 marks]

4. Plants can suffer from a disease called black spot. This causes black or purple spots on the upper surface of leaves. What causes black spot in plants? Tick **one** box.

☐ Bacteria   ☐ Viruses

☐ Fungi   ☐ Protists   [1 mark]

5. In recent years, rose black spot has become more common in urban gardens.

Untreated, black spot can quickly affect all the roses in a garden.

a State why the disease affects the plant's growth.

_____   [1 mark]

b Describe how the disease is transmitted.

_____

_____

_____

_____   [4 marks]

c A student is asked to explain four treatments for rose black spot.   [4 marks]

Worked Example

Since the disease is caused by fungus that produces spores, it is really important to make sure that the affected leaves and stems are removed immediately and burned. If the infected plant is allowed to remain untreated, the spores can be spread by rain or wind. In addition, fungicides can help kill the fungus.

This answer is worth 3 marks because the student has described three main points about immediate removal of stems and leaves and the need to burn them. The student also mentions fungicides. A further mark can be gained by showing an understanding that infected parts of the plant should not be composted as spores can survive and re-infect other rose plants.

Viral diseases

1. Viral diseases **cannot** be treated by antibiotics. Why? Tick **one** box.

☐ Viruses change their shape. ☐ Viruses are not living things.

☐ Viruses live inside cells. ☐ Viruses are too small. [1 mark]

2. HIV is a virus that can develop into AIDS. Describe **two** ways the HIV virus can be spread.

_____

_____ [2 marks]

3. Measles is a disease caused by a virus. Most children are given a vaccination against measles when they are young. Describe what the vaccination contains to make children immune to measles.

_____

_____ [1 mark]

Bacterial diseases

1. Which statement about bacteria is correct? Tick **one** box.

☐ All bacteria are harmful.

☐ Bacteria can infect plants and animals.

☐ Bacterial diseases are caused by poor diet.

☐ All bacteria have the same size and shape. [1 mark]

2. Salmonella bacteria cause food poisoning.

a How is *Salmonella* spread? Tick **all** the correct answers.

☐ sneezing into the air   ☐ sharing towels

☐ eating infected food   ☐ not washing hands before preparing food

☐ sexual contact   [2 marks]

b Explain why someone who eats infected food does not have symptoms immediately.

_____

_____

_____ [3 marks]

3. Give **two** symptoms of gonorrhoea.

1. _____

2. _____ [2 marks]

4. State **two** ways that the spread of gonorrhoea can be controlled.

1. _____

2. _____ [2 marks]

Malaria

1. Malaria is caused by which type of pathogen? Tick **one** box.

☐ Bacteria   ☐ Viruses   ☐ Fungi   ☐ Protists   [1 mark]

2. Malaria is caused by a protist called *Plasmodium*. The diagram shows stages in transmission of the malaria parasite by mosquitoes to humans.

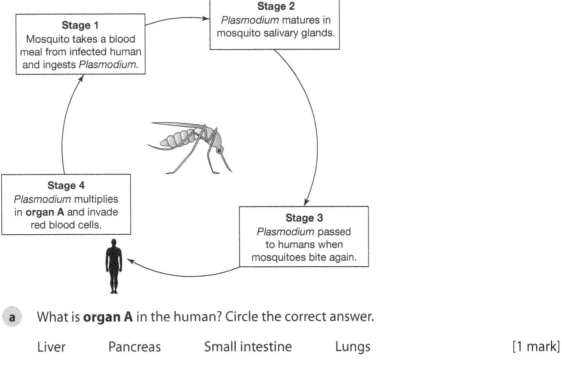

Stage 1
Mosquito takes a blood meal from infected human and ingests *Plasmodium*.

Stage 2
Plasmodium matures in mosquito salivary glands.

Stage 4
Plasmodium multiplies in **organ A** and invade red blood cells.

Stage 3
Plasmodium passed to humans when mosquitoes bite again.

a What is **organ A** in the human? Circle the correct answer.

Liver   Pancreas   Small intestine   Lungs   [1 mark]

b The plasmodia are transmitted by mosquitoes. What name is given to an organism that **spreads** disease rather than causing it? Tick **one** box.

☐ Virus   ☐ Vector   ☐ Fungus   ☐ Protist   [1 mark]

3. The mosquito has a complicated life cycle. Malaria is controlled by breaking the life cycle of mosquitoes in various ways.

Two ways to reduce malaria are:

1: draining stagnant water pools;

2: using mosquito nets.

Explain how these measures help to control malaria.

_____

_____

_____

_____   [4 marks]

Human defence systems

1. Draw **one** line from each part of the human defence system to its function.

White blood cells	Kills the majority of pathogens that enter via the mouth.
Stomach acid	Form scabs which seal the wound.
Platelets	Traps pathogens.
	Produce antimicrobial substances.

[3 marks]

2. The immune system has several methods of dealing with pathogens. The diagram shows how white blood cells attack pathogens by a process called phagocytosis.

Complete the diagram by drawing the 3rd stage of phagocytosis in the box provided.

Pathogen

1st stage 　　 2nd stage 　　 3rd stage

[2 marks]

3. Explain the adaptations of the respiratory system to protect against pathogens.

_____

_____

_____

_____ [4 marks]

Vaccination

1. In the sentences below, circle the correct underlined words or phrases to complete the sentences.

It is difficult to kill viruses inside the body because the virus: <u>is not affected by drugs</u> <u>/ lives inside cells / produces antitoxins</u>. The vaccine contains an <u>active / infective /</u> <u>inactive</u> form of the virus. The vaccine stimulates the white blood cells to produce <u>antibodies / antibiotics / drugs</u>, which destroy the virus. [3 marks]

2. In 1998, a scientist claimed that the mumps measles and rubella (MMR) vaccine caused autism in some children.

The graph shows the results of a study in Japan. It shows the number of children diagnosed with autism by age 7 and when the MMR vaccination programme began and ended.

> **Remember**
> With complex graphs like the one in Q2, it is easier to read the graph using the key. The bars show the percentage of children vaccinated. The line shows the number of children developing autism. This information is not clear from the axis labels alone.

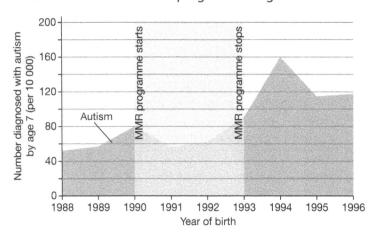

Maths **a** State how many children developed autism in the year the MMR vaccination started.

_____ [1 mark]

b Describe what happened to the number of children diagnosed with autism in the year after the vaccination programme was ended.

_____ [1 mark]

c Is there a link between MMR vaccination and autism? Explain your answer.

_____

_____ [2 marks]

3. Explain why new flu vaccines are made each year.

_____

_____

_____

_____ [4 marks]

Antibiotics and painkillers

....

1. Medicines contain useful drugs that relieve the symptoms caused by pathogens or kill the pathogens. Draw **one** line from each medicine to its correct function.

Kill viruses without damaging human cells.

Antibiotics

Relieve symptoms of infection.

Painkillers

Kill bacteria by interfering with the process that makes bacterial cell walls.

[2 marks]

2. Read the passage about the use of antibiotics to treat MRSA.

> The emergence of strains of bacteria resistant to antibiotics is of great global concern.
>
> MRSA (Meticillin-resistant _Staphylococcus aureus_) can cause a range of illnesses from skin disorders to deadly diseases like meningitis and pneumonia.
>
> MRSA is a 'superbug' because it has become resistant to the antibiotic meticillin.

Worked Example

a Explain how a population of antibiotic-resistant bacteria such as MRSA can develop from non-resistant bacteria. [3 marks]

Antibiotic will only kill non-resistant bacteria. If some bacteria survive, they will produce offspring thus increasing the population of resistant bacteria.

This response is only worthy of 2 marks. The student has an idea that some bacteria can survive the antibiotic – this gains 1 mark. The student also knows that the resistant bacteria breed and produce more resistant bacteria. However, the student should also have written about how the non-resistant bacteria become resistant – through mutation or variation.

Synoptic **b** Describe **two** ways of slowing down the rate of development of resistant strains of bacteria.

1. _____

2. _____ [2 marks]

3. Why will antibiotics **not** get rid of flu?

_____

_____ [2 marks]

Remember
Q3 requires you to apply your knowledge that 1) antibiotics affect only bacteria and 2) flu is caused by a virus.

4. Explain the limitations of antibiotics.

_____

_____

_____ [3 marks]

Making and testing new drugs

1. Why must medical drugs be tested before they are used on patients?

Tick **all** boxes that apply.

☐ To check they work efficiently.

☐ To check they are safe to use.

☐ To make them as cheap as possible.

☐ To find the right dose.

☐ To check if they will work on animals. [2 marks]

Common misconception
On questions that ask you to 'tick all those that apply', do **not** assume that the marks awarded show how many ticks are needed. In this case, there are three ticks required but only **2** marks awarded.

2. During clinical trials, new drugs are tested on animals and humans.

What would the new drug have been tested on before animals and humans?

_____ [1 mark]

3. Researchers sometimes use traditional medicines when starting to develop new drugs.

Draw **one** line from each medicine to match it with its correct source.

Heart drug digitalis	Willow trees
Painkiller aspirin	Foxgloves
Antibiotic penicillin	Tree bark
Anti-malarial quinine	Mould

[2 marks]

Photosynthesis reaction

1. Use the words from the box to complete the sentences.

| light | heat | leaves | stems | mitochondria | chemical | chloroplasts |

Photosynthesis is a chemical reaction which happens in the

_____ of green plants.

During photosynthesis _____
energy is absorbed by chlorophyll, a green substance found

in _____ in some plant and
algae cells. [3 marks]

Common misconception
Many people think plants get their food from the soil, but this is a misconception. A plant makes its own 'food' in the form of glucose by the process of photosynthesis. Plants have evolved to harvest energy from the Sun to produce their own supply of glucose.

2. Complete the word equation for photosynthesis.

light
_____ + water $\longrightarrow$ glucose + _____ [2 marks]

3. Photosynthesis takes in energy from light during photosynthesis. What term **best** describes a reaction that takes in energy? Tick **one** box.

☐ Exothermic   ☐ Endothermic   ☐ Physical   ☐ Chemical   [1 mark]

4. Name the chemical reaction used by plants to obtain energy from glucose.

_____ [1 mark]

5. Why is photosynthesis the first step in 'making food for every animal on the planet'?

_____

_____

_____ [2 marks]

Rate of photosynthesis

1. List **three** factors that limit the rate of photosynthesis.

_____ [3 marks]

2. Elena investigates the rate of photosynthesis in *Cabomba* pondweed.

Practical She has this apparatus:

- 7 cm-long pieces of pondweed
- beaker
- funnel
- lamp

- metre ruler
- stop clock
- 1% sodium hydrogen carbonate (or water)

a *Cabomba* produces bubbles of oxygen as it photosynthesises. Using this apparatus, describe a method Elena could use to investigate the effects of light intensity on the rate of photosynthesis in pondweed.

Your answer should include:

- what you would measure
- what variables you would control.

_____

_____

_____

_____

_____

_____ [6 marks]

Maths **b** Elena's results are shown in the table.

Distance from lamp [cm]	Number of bubbles produced in a minute			
	Test 1	Test 2	Test 3	Mean
10	46	39	42	
15	34	31	29	
20	21	23	24	

Complete the missing values in the table by calculating the mean number of bubbles produced for each distance from the lamp. Give your answers to 3 significant figures.

_____ [3 marks]

  c Give **two** reasons why counting bubbles is **not** the most accurate way of measuring the amount of oxygen given off.

_____

_____ [2 marks]

  d Write a conclusion about the effect of light intensity on the rate of photosynthesis. Use the information in the table.

_____ [1 mark]

3.

Higher Tier only

The inverse square law allows you to calculate how much light falls on pondweed at different distances from the light source. It is calculated using this formula:

$$\text{Light intensity} = \frac{1}{\text{distance}^2}$$

Distance from lamp (cm)	Light intensity (arbitrary units)
10	0.01
15	0.004
20	_____

Use the formula to calculate the light intensity for 20 cm. Write your answer in the table.

_____ [1 mark]

4.

Higher Tier only

Light intensity obeys the inverse square law. This means that if you double the distance you:

Tick **one** box.

☐ Quarter the intensity   ☐ Halve the intensity

☐ Double the intensity   ☐ Minus the intensity   [1 mark]

Limiting factors

1.

Higher Tier only

A scientist monitors the concentration of dissolved oxygen in a pond over a 24-hour period. She finds that the dissolved oxygen concentration is always highest at about 4 pm. Explain her findings.

_____

_____ [2 marks]

2.

Higher Tier only

Margery want to speed up the growth of her tomato plants. She puts them in a greenhouse with a paraffin heater. List **two** ways this could speed up the growth.

_____

_____ [2 marks]

3.

Worked Example

The graph shows how carbon dioxide concentration affects the rate of photosynthesis.

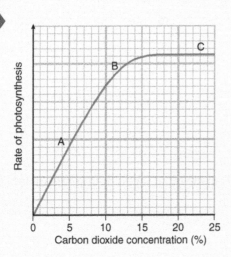

> **Remember**
>
> For the exam you will be expected to extract information and interpret graphs about rates of photosynthesis and limiting factors. Make sure you can explain 'rate of photosynthesis' graphs for carbon dioxide, light and temperature.

a Describe and explain the shape of the curve between points A, B and C. [4 marks]

A student's answer to this question is given here.

Between A and B the rate of photosynthesis increases linearly with increasing carbon dioxide concentration.

> The student has used the term 'linear', which accurately describes the relationship.

Between B and C gradually the rate decreases, and at a certain carbon dioxide concentration the rate of photosynthesis becomes constant.

At point C a rise in carbon dioxide concentration has no effect on the rate of photosynthesis.

> It is good practice to use actual figures from the graph in a description. It would be better to say 'at 15% carbon dioxide concentration'.

> Although the student has written a description, they have not explained what they have described so they would only get 2 of the 4 marks. They would need to explain the description by saying that between A and B the rate of photosynthesis increases with carbon dioxide concentration, because carbon dioxide is needed for photosynthesis, so the more there is, the faster the rate. Between B and C, the rate does not increase because other factors such as light intensity become limiting.

b The concentration of carbon dioxide in the air is 0.04%. Use information in the graph to explain whether a gardener could make plants in his greenhouse grow faster by giving them extra carbon dioxide.

Higher Tier only

_____

_____ [2 marks]

Uses of glucose from photosynthesis

1. Use the words from the box to complete the sentences.

starch	glycogen	storage	respiration	insulation	growth	energy

The glucose produced by a plant during photosynthesis is converted into

_____, fats and oils for _____.

Some glucose is used to make cellulose for cell walls, and proteins for

_____ and repair. It is also used by the plant to

release energy by _____. [4 marks]

2. Which of the following statements are true? Tick **two** boxes.

☐ All plant cells carry out photosynthesis in the day and then switch to respiration at night.

☐ Plant cells can photosynthesise at day or night depending on when they need to produce glucose.

☐ Plant cells respire all the time, but some also carry out photosynthesis when light is available.

☐ Only some plant cells can carry out photosynthesis; some, such as root hair cells, do not. [2 marks]

3. The part of the potato plant we eat is the tuber. It is an enlarged underground stem that stores starch for the plant. New potato plants grow from potato tubers.

Suggest how new potato plants obtain energy needed for growth.

_____

_____ [1 mark]

4. Explain why the potato plant no longer needs this energy source once it has grown above the soil.

_____

_____ [1 mark]

5. Why do the tubers store starch, **not** glucose?

_____ [1 mark]

6. Explain why a plant that does **not** absorb enough nitrate ions might have stunted growth.

_____

_____

_____ [2 marks]

Cell respiration

1. Use the words from the box to complete the sentences.

| glucose | cells | aerobically | anaerobically | nuclei | mitochondria | oxygen |

Respiration is a series of reactions in which energy is released from the reactant, which is

_____. Respiration can take place _____

(uses oxygen) or _____ (without oxygen) in the form of respiration

that uses oxygen. Aerobic respiration happens inside the _____
found in cells. [4 marks]

2. Which of the following statements about respiration is true? Tick **one** box.

☐ Respiration is another name for breathing.

☐ Glucose is produced during respiration.

☐ Respiration does not happen when the body is at rest.

☐ Respiration releases energy from glucose. [1 mark]

3. Complete the word equation for aerobic respiration:

glucose + _____ ⟶ carbon dioxide + _____ (+ energy) [2 marks]

4. Give **two** ways an organism uses the energy released from respiration.

_____ [2 marks]

5. Explain why respiration is an exothermic reaction.

_____ [1 mark]

6. Some of the energy transferred by respiration in our cells is used to synthesise new molecules that the body needs. Give **two** examples of smaller molecules that are linked together to form larger molecules using energy from respiration.

_____

_____ [2 marks]

7. Long-distance runners often eat meals containing a lot of carbohydrate over three days before a race. How does this help muscles to work well during a race? [2 marks]

Worked Example

A student's answer to this question is given here.

The carbohydrate provides the glucose needed for respiration.

This is correct and would gain 1 mark. But to get 2 marks, you would need to say that the runner's muscle cells will use lots of energy to contract during the race, so a good supply of glucose will enable the muscles to respire at a faster rate.

Anaerobic respiration

1. What does the word **anaerobic** mean? Tick **one** box. [1 mark]

☐ With oxygen ☐ Without oxygen ☐ No energy ☐ Without carbon dioxide

2. What is the product of anaerobic respiration in muscle cells?

_____ [1 mark]

3. Explain why our cells normally respire aerobically, rather than anaerobically.

_____

_____ [2 marks]

4. Oliver is sprinting away from a bull. Explain what type of respiration takes place in his muscle cells when he begins to sprint. What type of respiration occurs after sprinting for several minutes?

_____

_____

_____ [3 marks]

5. Complete the word equation to show anaerobic respiration in plant and yeast cells.

glucose $\longrightarrow$ _____ + _____ [2 marks]

6. Anaerobic respiration in yeast cells is called fermentation.

Give **two** uses of fermentation.

_____ [2 marks]

7. Compare anaerobic respiration in a yeast cell with anaerobic respiration in a muscle cell.

_____

_____

_____ [3 marks]

Command words

The command word 'compare' requires a description of the similarities and/or differences between things. Make sure you do not just write about one.

Response to exercise

1. Use the words in the box to complete the sentences.

| deoxygenated   breathing volume   oxygenated   glucose |
| carbon dioxide   carbon monoxide   energy |

When you exercise, your heart rate, breathing rate and _____

increase to supply the muscles with more _____ blood.

The blood delivers oxygen and _____ to the respiring muscle

cells and takes away _____. [4 marks]

2. Which statement **best** describes what happens when muscles become fatigued during vigorous activity? Tick **one** box.

☐ They work more efficiently.   ☐ They work less efficiently.

☐ They stop contracting.   ☐ They produce urea.   [1 mark]

3. Explain why a person continues to breathe heavily after strenuous exercise has stopped.

_____

_____ [2 marks]

4. The graph shows how oxygen uptake by the lungs changes during exercise and recovery.

Higher Tier only

Which area shows:

a The amount of oxygen absorbed by the lungs during exercise?

_____ [1 mark]

b The amount of oxygen needed for aerobic respiration during exercise?

_____ [1 mark]

c The oxygen debt?

_____ [1 mark]

d Explain why areas A and C are the same size.

_____ [1 mark]

Homeostasis

1. Which of the following are controlled by homeostasis?

Tick **two** boxes.

☐ Blood glucose levels   ☐ White blood cells   ☐ Body mass

☐ Body temperature   ☐ Blood oxygen levels   [2 marks]

2. Name the part of the body where the thermoregulatory centre is found.

_____ [1 mark]

3. Describe the function of the thermoregulatory centre.

_____ [1 mark]

4. Read the statements and circle the **correct** response.

Hormones travel around the body in the form of electrical impulses.	TRUE / FALSE
Organs that secrete hormones are called glands.	TRUE / FALSE
Hormones are carried all over the body but only affect target organs.	TRUE / FALSE
There are three control systems responsible for homeostasis.	TRUE / FALSE
Effectors that bring about a response can be muscles or glands.	TRUE / FALSE

[5 marks]

5. There are **three** organs of the body involved in excretion – what are they and what do they excrete?

Complete the table below.

Organ	Substance excreted

[3 marks]

The nervous system and reflexes

1. A man accidentally touches a hot plate. His hand immediately moves away. The diagram shows the structures involved in this response.

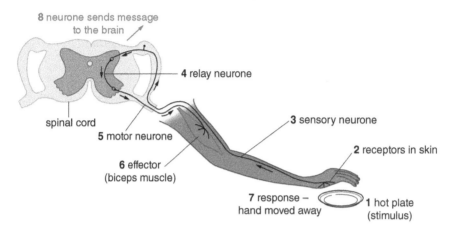

Use the correct word or phrase from the diagram to complete each sentence.

The stimulus is detected by the _____ in the skin. Impulses

travel along a _____ to the central nervous system.

A _____ neurone carries the impulse to the motor neurone.

The hand is then moved away from the hot plate by the _____

which in this example is the biceps muscle. [4 marks]

2. The distance travelled by the impulse in the reflex action shown in the diagram in question 1 is 1.26 metres.

   The speed of the nerve impulse is 90 metres per second.

   Maths
   To find the time taken, divide the length by the speed of the nerve impulse. The unit for speed is metres per second; therefore, the unit for time is seconds.

   a   Use this information to calculate the time taken for this reflex action.

   Show your working out clearly.

   _____

   Time taken for reflex action = _____ seconds   [1 mark]

b Suggest why the reflex action might take longer than your answer to part **a**.

_____ [1 mark]

3.

Practical

Two students are investigating reaction times:

1. Student A sits with an arm resting on the edge of a table.

2. A ruler is held vertically between the thumb and forefinger of Student B with the bottom of the ruler level with the thumb.

3. Student B drops the ruler.

4. Student A catches the ruler as fast as they can. The distance fallen by the ruler is recorded.

Describe **two** ways the students can ensure this method produces results that are repeatable.

Student B

Student A

_____

_____

_____ [2 marks]

4.

Practical

A scientist measured reaction time using a computer program.

This is the method used.

1. The computer shows a blank yellow screen.

2. When the screen changes from yellow to blue the person being tested presses a button as quickly as possible.

3. This is repeated six times and then a mean reaction time is calculated.

Explain why using a computer program to measure reaction times is likely to produce more repeatable results than the ruler method in question 3.

_____

_____ [2 marks]

Hormones and the endocrine system

1. Hormones are chemicals that can pass information around the body.

a Describe how hormones travel around the body.

_____ [1 mark]

b State the name given to organs that secrete hormones.

_____ [1 mark]

2. Write the names of glands **A** and **B** on the diagram.

[2 marks]

A _____

B _____

3. **a** Identify the master gland from ② above.

Master gland = _____ [1 mark]

b Why is it called the master gland?

_____

_____ [1 mark]

4. **a** Complete the table below to show which hormone is secreted by which gland.

Gland / organ	Hormone
_____	Follicle stimulating hormone (FSH)
Thyroid	_____
_____	Adrenaline
Ovaries	_____

[4 marks]

b The pituitary gland produces hormones that have a direct effect on their target organs. Some pituitary hormones have an indirect effect – they cause other glands to secrete hormones.

Describe **two** hormones produced by the pituitary and their effects on their organs. [6 marks]

A student's answer to this question is given below.

The pituitary gland produces thyroid-stimulating hormone which acts on the thyroid gland. It also produces the hormone FSH that makes the ovaries release oestrogen.

> This answer gains 5 marks out of a possible 6; 2 marks for correctly stating the hormone the pituitary gland produces and the target organ for that hormone. However, 1 mark was lost because there is no mention of what the target organ does after it is stimulated by the hormone. The second sentence gains full credit as the student has stated correctly the name of the hormone, the target organ and the response.

Controlling blood glucose

1. A doctor monitors the blood glucose of two patients.

The patients were both given a glucose drink at time 0.

The graph shows the effect of the drink on the glucose levels in both patients.

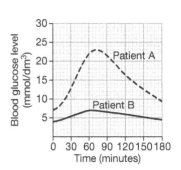

a Identify which patient has diabetes.

_____ [1 mark]

b Describe **one** other symptom of diabetes that this patient might have.

_____ [1 mark]

c Explain why people with Type 1 diabetes need to have insulin injections.

_____

_____ [2 marks]

2. Explain what happens if **too much** insulin is taken.

_____

_____ [2 marks]

3. There are two types of diabetes called Type 1 and Type 2.

Complete the table comparing Type 1 and Type 2 diabetes.

	Type 1 diabetes	Type 2 diabetes
Who does it affect?	a Children, teenagers and young adults	b
Caused by ...	c Infection of the pancreas, genetic link	d
Reason	e	f Body no longer responds to its insulin
Treatment	g Insulin injections and a healthy diet	h

[4 marks]

4. How does the body of a healthy person restore blood sugar levels if the level drops too low?

Higher Tier only

_____

_____

_____ [3 marks]

Hormones in human reproduction

1. Complete the table below.

Hormone	Organ where it is produced	Function of hormone
Oestrogen	_____	Female reproductive hormone
_____	Testes	Male reproductive hormone

[2 marks]

2. Hormones are responsible for regulating the female reproductive organs. Complete the sentences below.

The monthly release of an egg from the _____ is controlled by hormones.

Hormones also control the thickness of the _____ lining. Hormones

given to women to stimulate the release of eggs are called _____. [3 marks]

3. Describe the function of FSH and LH in the menstrual cycle.

_____

_____ [2 marks]

Hormones interacting in human reproduction

1. In humans, fertilisation can only occur within a few days of a woman ovulating.

Higher Tier only The graphs show hormone levels during the menstrual cycle.

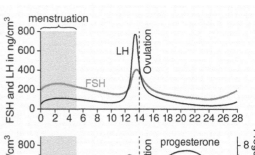

a Which hormone stimulates ovulation? Tick **one** box.

☐ Oestrogen ☐ Progesterone

☐ FSH ☐ LH

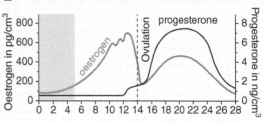

[1 mark]

b Which of the following is caused by high levels of oestrogen? Tick **one** box.

☐ FSH production is stimulated. ☐ Progesterone production is inhibited.

☐ Menstruation occurs. ☐ LH production is stimulated. [1 mark]

c Explain the reasons why the uterus lining is maintained if fertilisation occurs.

_____

_____ [2 marks]

2.

Higher Tier only

The diagram shows how the depth of the uterus lining (endometrium) changes during the menstrual cycle.

Use the diagram to explain the changes in depth of the uterus lining and the hormones involved.

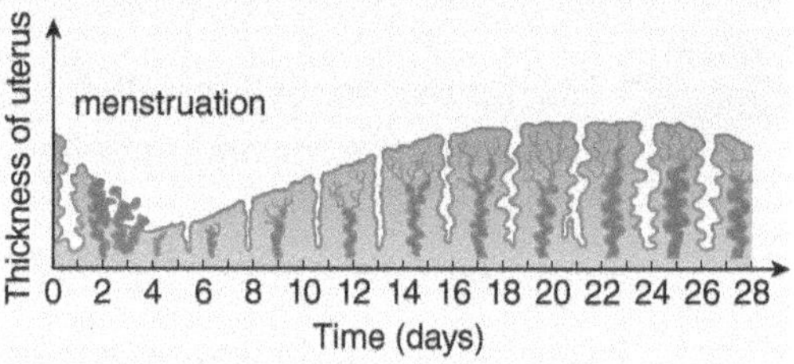

_____

_____

_____

_____

_____

_____ [6 marks]

Contraception

1. The table below shows some methods of contraception.

Type of contraception	Percentage (%) of pregnancies prevented
Contraceptive injection	94
Combined pill	91
Diaphragm	85
Female sterilisation	99

a State which method of contraception shown in the table is **most** effective at preventing pregnancies.

_____ [1 mark]

b State which method offers **best** protection against sexually transmitted diseases.

_____ [1 mark]

There are two types of intrauterine device (IUD), the plastic IUD and the copper IUD.

Both types are 99% effective.

The table gives information about both types of IUD.

	Plastic IUD	Copper IUD
How long contraceptive is effective	5 years	5–10 years
How it works	Progesterone is released which thickens mucus around cervix preventing passage of sperm.	Copper acts as a spermicide by changing the fluids inside uterus which prevents sperm surviving.
Other advantages	Can decrease period cramping.	Can be used as emergency contraception.

Possible side effects	Cramping or back ache for a few days after IUD is inserted. Bleeding between periods for first 3 months.	Heavy and painful periods. Bleeding between periods.

c Evaluate the use of the plastic IUD and the copper IUD as a contraceptive.

_____

_____

_____

_____

_____

_____ [5 marks]

Command words

The command word 'evaluate' means to judge from available evidence. An evaluation goes further than a 'compare question'. You will need to write down some of the points for and against both types of IUD to develop an argument.

2. Below are some facts about using birth control pills.

Which facts show the advantages of using birth control pills? Tick **three** boxes.

☐ Birth control pills are 99% effective in preventing pregnancy.

☐ The hormones in the pills have some rare but serious side effects.

☐ This method of birth control gives no protection against sexually transmitted diseases.

☐ The hormones in the pills give protection against some women's diseases.

☐ The woman has to remember to take the pill every day.

☐ The woman's monthly periods become more regular. [3 marks]

Using hormones to treat infertility

1. For women who are **unable** to conceive, hormones can be given as a fertility drug.

Higher Tier only

Which hormones can be used to treat infertility? Tick **two** boxes.

☐ Adrenaline   ☐ Thyroxine   ☐ FSH

☐ Oestrogen   ☐ LH   ☐ Testosterone   [2 marks]

2. These are some advantages and disadvantages of using fertility and contraceptive drugs.

Higher Tier only

What are the **disadvantages** of using fertility and contraceptive drugs? Tick **three** boxes.

☐ Prevent unwanted pregnancy.

☐ May increase chance of getting a sexually transmitted disease.

☐ May cause side-effects in female body.

☐ Regulate the menstrual cycle.

☐ Prolonged use may prevent later ovulation.

☐ Can stimulate egg release.

☐ May cause multiple births.   [3 marks]

3. Read the information about fertility treatment.

Higher Tier only

In vitro fertilisation (IVF) is a fertility treatment. It involves giving a woman hormones to stimulate maturation of eggs. The eggs mature in the woman's ovary before being fertilised by sperm in a laboratory.

The fertility drugs used in IVF can cause ovarian hyperstimulation syndrome, which can be life threating.

In vitro maturation (IVM) is a different fertility treatment. Immature eggs are collected from the ovaries, matured in a laboratory and then injected with sperm.

It does not use fertility drugs so women are not at risk of ovarian hyperstimulation syndrome.

IVF costs about £5000 per round of treatment. IVM costs about £3000.

IVF has a success rate of around 40%. IVM has a success rate of around 20%.

Evaluate the use of IVF and IVM to treat infertility.

_____

_____

_____

_____ [4 marks]

Negative feedback

1.

Human body temperature is 37°C. When a person is in a hot environment where the air temperature is much higher than 37°C, changes take place to make sure their body temperature remains at 37°C.

a Explain **two** changes that take place in the body to keep its temperature at 37°C in a hot environment.

_____

_____ [2 marks]

b Homeostasis involves negative feedback.

With reference to body temperature, describe what is meant by the term 'negative feedback'.

_____

_____

_____ [3 marks]

2.

The graph below shows how a person's blood glucose levels change after eating a meal.

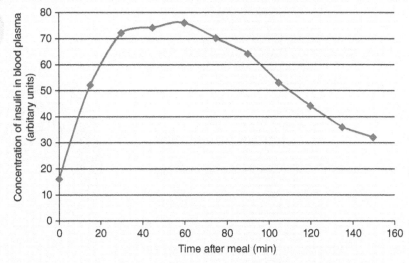

Explain how the graph shows an example of negative feedback.

_____

_____

_____

_____

_____

_____

_____ [4 marks]

Sexual reproduction and fertilisation

1. A lion and a tiger can breed together. The offspring is known as a liger.

a Which technique is used to produce a liger?
Tick **one** box.

☐ Cloning ☐ Sexual reproduction

☐ Asexual reproduction ☐ Genetic engineering

[1 mark]

b A liger has features of both the lion and the tiger. Explain why.

_____

_____

_____ [2 marks]

2. The diagram shows some of the stages of reproduction in cats.

a Name the type of reproduction shown here.

_____ [1 mark]

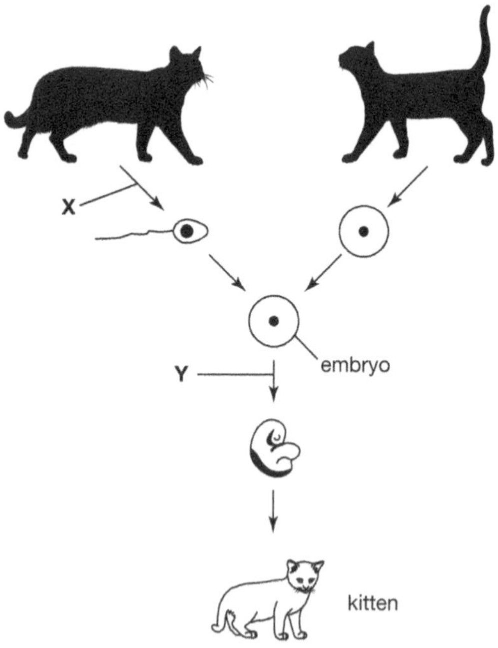

b Circle a part of the diagram that shows a
gamete. [1 mark]

c State the letter on the diagram showing:

Mitosis: _____ [1 mark]

Meiosis: _____ [1 mark]

X

Y

embryo

kitten

d The kitten grows into a healthy adult cat.
Explain why the cat looks similar, but not
identical to either parent.

_____

_____ [2 marks]

Asexual reproduction

1. Strawberry plants grow runners. The runners sprout from the parent plant. As the runners grow out across the ground, they attempt to grow roots into the soil. If a rooting attempt is successful, a new plant crown will grow above the roots.

Use the words from the box to complete the following sentences.

asexual	chromosome	clone	gamete	gene	sexual variant

The production of runners in strawberry plants is an example of _____ reproduction. The does not involve the production of a _____.

The new strawberry plant that grows from a runner is called a _____. [3 marks]

2. Sexual reproduction in plants and animals involves the fusion of gametes.

Name the gametes in:

a Plants

_____ [1 mark]

b Animals

_____ [1 mark]

3. Potatoes can reproduce by asexual and sexual reproduction.

Explain the differences between potatoes produced by asexual and sexual reproduction.

_____

_____

_____ [2 marks]

Cell division by meiosis

1. In which of the following would you expect meiosis to take place? Tick **two** boxes.

☐ Ovaries ☐ Uterus ☐ Bladder ☐ Testes ☐ Prostate [2 marks]

2. An organism has 28 chromosomes in each of its body cells. State how many of these chromosomes will have come from the female parent.

> **Remember**
> Half of the chromosomes come from each parent.

_____ [1 mark]

3. The diagram shows a simplified model of chromosomes in a cell dividing by meiosis.

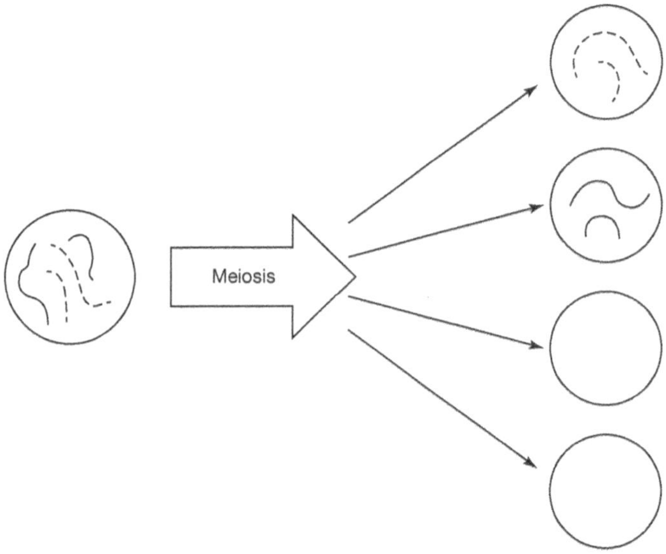

Meiosis

a Complete the two empty cells shown in the diagram to show the rearranged chromosomes in two genetically different cells. [2 marks]

b Name the structure in the cell where chromosomes are found.

_____ [1 mark]

c Name the structure in a female where meiosis takes place. _____ [1 mark]

d Compare mitosis and meiosis in terms of the number of chromosomes passed on to each daughter cell.

_____

_____ [2 marks]

DNA, genes and the genome

. . . .

1. Choose words from the box to complete the sentences below.

| body   chromosomes   clones   cytoplasm   genes   nucleus   sex |

Information is passed from parents to their young, in _____ cells.

Each characteristic, e.g. fur colour, is controlled by _____. The structures that

carry information for a large number of characteristics are called _____. The

part of the cell which contains these structures is called the _____.   [4 marks]

2. Circle the correct answer.

In the nucleus of a cell, genes are part of:

chromosomes   membranes   receptors   cytoplasm   [1 mark]

3. The Human Genome Project was a study to map all the genetic information on the chromosomes of a human being.

Explain **two** benefits of understanding a person's genome.

_____

_____   [2 marks]

4. DNA is made of bases.

   a What are the bases that make up DNA?

   _____ [1 mark]

   b How do these bases pair up?

   _____ [1 mark]

5. Give **three** ethical issues associated with the Human Genome Project.

   1. _____

   2. _____

   3. _____ [3 marks]

Inherited characteristics

1. Put a ring around the correct answer to complete each sentence below.

   The alleles present in the cell are the <u>genotype/phenotype</u>.

   A recessive allele is <u>always expressed/only expressed</u> if two copies are present.

   If **two** alleles present are recessive, the organism is <u>heterozygous/homozygous</u>.   [3 marks]

2. Fur colour in mice is controlled by one pair of genes. The allele for black fur (B) is dominant to the allele for brown fur (b). A black coat mouse is crossed with a brown furred coat. They produce a litter of eight mice. Four of the mice have black fur and four of the mice have brown fur.

   Complete the diagram to show how the mice inherited their fur colour.

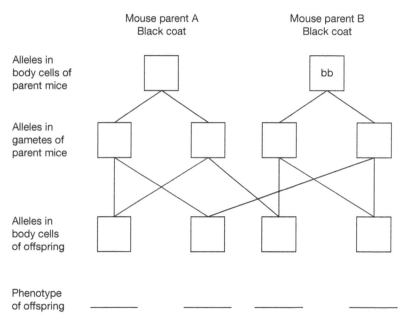

   Use the symbols N and n for the alleles.

   The alleles of the brown parent have been inserted for you.   [3 marks]

3. Pea plants that are tall have the genotype TT or Tt. Pea plants that are short have the genotype tt.

a Draw a Punnett square to show the genotypes of the offspring produced by a cross between two plants that are both Tt. [3 marks]

Remember

Place the alleles of one parent of the left hand side of the square and the alleles of the other parent along the top of the square. Put one allele next to each box. Use capital letters for the dominant genes and lower-case letters for the recessive alleles.

b Why might your prediction of fur colour in the F1 generation **not** be proved right?

_____ [1 mark]

c Using the example of the mice coat colour to help:

(i) describe the difference between dominant and recessive alleles.

_____

_____ [2 marks]

(ii) describe the difference between alleles and genes.

_____

_____ [2 marks]

(iii) describe the difference between homozygous and heterozygous chromosomes.

_____

_____ [2 marks]

Inherited disorders

1. Haemophilia is a genetic disorder which is sex-linked.

The diagram shows a family tree.

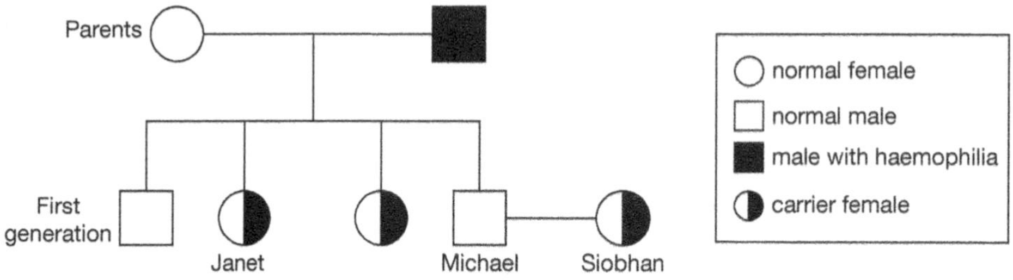

XH represents a normal X chromosome.

Xh represents an X chromosome carrying the haemophilia allele.

Y represents a Y chromosome.

a Give the genotype of the parents.

Father: _____ **Mother:** _____ [2 marks]

b Give the phenotype and genotype of Janet.

Phenotype: _____ **Genotype:** _____ [2 marks]

c Explain how Michael does **not** have haemophilia even though his father does.

_____

_____ [2 marks]

Higher Tier only

d Michael and Siobhan have children. What proportion of the children will have

haemophilia? _____ [1 mark]

Maths

e Why are there fewer females with haemophilia than males?

_____

_____ [2 marks]

2. Cystic fibrosis is an inherited disorder caused by a mutated allele.

The allele that causes cystic fibrosis is a recessive allele.

a A couple are having a baby. Neither parent has cystic fibrosis. [2 marks]

Complete the Punnett square to show the genotypes of the offspring.

Use F for the dominant allele for cystic fibrosis and f for the recessive allele for cystic fibrosis.

Put a circle around any children with cystic fibrosis.

Mother

	F	f
Father F		
f		

b Determine the probability of the couple having a child with cystic fibrosis.

Probability: _____ [1 mark]

c Which of the following terms **best** describes the genotype Ff? Tick **one** box.

☐ Heterozygous   ☐ Homozygous dominant

☐ Homozygous recessive   ☐ Homozygous   [1 mark]

3. Alex has a condition called **polydactyly**. The condition causes extra fingers to form on the hands or extra toes on the feet. The condition is caused by a dominant gene (P).

a Alex's mother does not have polydactyly. Will Alex's father have the condition? Explain why.

_____

_____ [2 marks]

b Alex's father is heterozygous for the condition.

Complete the genetic diagram to show the probability of Alex's brothers or sisters inheriting the condition.

		Father	
Mother			

Probability of offspring having polydactyly is _____ [4 marks]

Sex chromosomes

1. The diagram shows how sex is determined in offspring.

    a Circle the part of the diagram that represents an allele carried by an egg cell. [1 mark]

    b Write the genotype of the offspring that are male.

    _____ [1 mark]

	Mother		
		X	X
Father	X	XX	XX
	Y	XY	XY

2. A couple have three children. All their children are girls. The couple are expecting their fourth child.

State the probability that the child will be a boy.

Probability = _____ [1 mark]

3. The diagram below shows sex inheritance.

    a Complete the missing alleles to show how sex is determined.

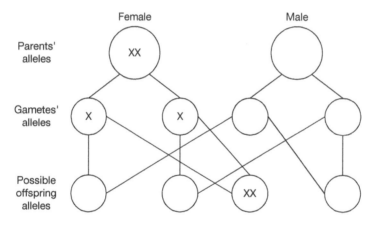

[3 marks]

A couple already have three boys.

    b Determine the probability that a child born from the diagram will be a girl.

Probability: _____ [1 mark]

Variation

1. A gardener takes cuttings from a plant. Cuttings produce clones of the parent plant. The gardener plants the cuttings at different distances from a tall fence.

The diagram shows the variation in the growth of the cuttings after five weeks.

a Define the term variation.

_____ [1 mark]

b Give **two** possible reasons the plants showed variation after five weeks.

1. _____

2. _____ [2 marks]

c Which type of variation is shown by the plants after five weeks? Tick **one** box.

☐ Genetic ☐ Environmental [1 mark]

☐ Genetic and environmental ☐ It is impossible to tell

d Give a reason for your answer to **c**.

_____ [1 mark]

2. A population of bacteria that have resistance to an antibiotic develop from a population of bacteria that do **not** have resistance.

Explain how this variation in the population of bacteria occurs.

_____

_____

_____

_____ [3 marks]

3. **a** Give **two** factors which bring about variation in humans.

1. _____

2. _____ [2 marks]

b There is large variation within a population of a species.

A student was asked to suggest **two** factors which could produce this range of variation. [2 marks]

Worked Example

Variation is because of the genes that are inherited or the environmental conditions.

This answer only gets 1 mark for correctly pointing out both genetic and environmental factors. The student could have gained an extra mark by writing about mutations causing variation in the population. Another credit-worthy suggestion is that there is a large gene pool for this population, which means a large number of combinations of genes are possible.

Evolution by natural selection

1. Draw a circle around the correct answer to complete each sentence. [2 marks]

Evolution can be explained by a theory called genetic engineering/mutation/natural selection.

This theory states that all living things have evolved from monkeys/dinosaurs/simple life forms.

2. Tick **one** box to complete each sentence.

a Charles Darwin suggested the theory of evolution by _____ selection.

☐ artificial   ☐ natural   ☐ asexual   [1 mark]

b Most scientists believe that life first developed about _____ years ago. Tick **one** box.

☐ three billion   ☐ three million   ☐ three thousand   [1 mark]

3. Describe how evolution occurs by the process of natural selection.

_____

_____

_____

_____ [4 marks]

4. Populations of animals can become isolated (split up in different regions) when people build new roads or cities. When populations of the same species are isolated in this way, new species are more likely to develop. Explain why this may happen.

_____

_____

_____

_____ [4 marks]

Fossil evidence for evolution

1. Scientists have used fossils to give an indication of the species that existed in the past. The diagram shows how the number of species of different vertebrates has changed over time.

The width of each band is proportional to the number of species that existed at the time.

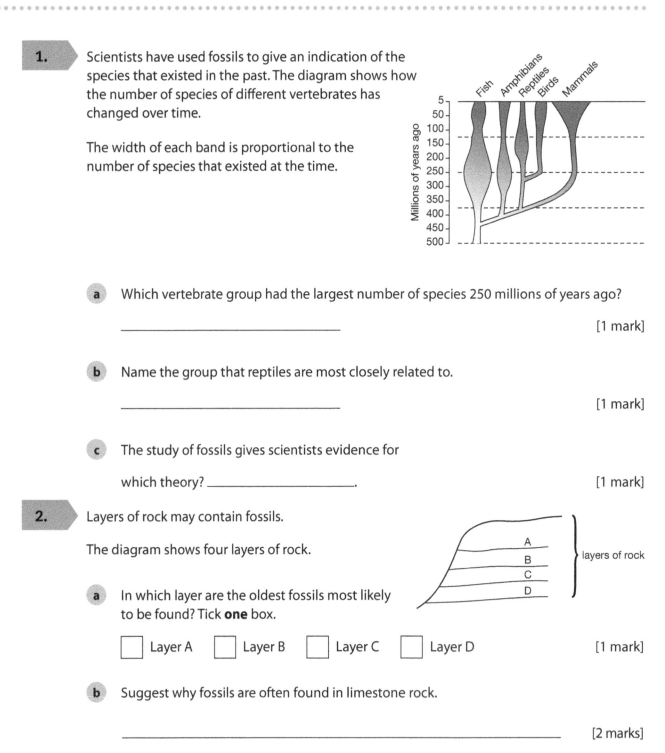

a Which vertebrate group had the largest number of species 250 millions of years ago?

_____ [1 mark]

b Name the group that reptiles are most closely related to.

_____ [1 mark]

c The study of fossils gives scientists evidence for

which theory? _____. [1 mark]

2. Layers of rock may contain fossils.

The diagram shows four layers of rock.

a In which layer are the oldest fossils most likely to be found? Tick **one** box.

☐ Layer A ☐ Layer B ☐ Layer C ☐ Layer D   [1 mark]

b Suggest why fossils are often found in limestone rock.

_____ [2 marks]

Other evidence for evolution

1. The cultivated banana is sterile and seedless. It reproduces asexually, producing shoots or suckers that develop into new plants. It is difficult to develop disease-resistant varieties of this plant.

In the 1950s, the main variety grown in banana plantations was Gros Michel. This was wiped out by Panama disease, caused by a soil fungus (*Fusarium oxysporum*). The disease spread rapidly from one plant to another and from plantation to plantation.

Following the devastation, a variety resistant to Panama disease, Cavendish, was planted and has been used ever since. However, Cavendish was susceptible to the disease Sigatoka, caused by another fungus (*Mycosphaerella fijiensis*). Only massive amounts of fungicide spray keep Sigatoka under control. It seems that as soon as a new fungicide is used, the fungus develops resistance to it.

Synoptic **a** State the genus of the fungus that causes Panama disease.

_____ [1 mark]

b Explain why Panama disease spread so quickly and wiped out Gros Michel.

_____

_____ [3 marks]

> **Remember**
> This answer requires you to notice that plants in the plantation grow close together; also the plants are clones of each other and therefore susceptible to the same diseases.

c Explain how the fungus that causes Sigatoka becomes resistant to fungicides.

_____

_____

_____

_____

_____ [4 marks]

2. Explain how evidence from antibiotic resistance in bacteria supports the theory of evolution. [6 marks]

A student has written the following answer to this question.

Bacteria evolves to get a higher tolerance or resistance to antibiotic. Therefore, antibiotic doesn't really affect bacteria. As the bacteria multiply they have a wide range of genetic variation. The bacteria carrying resistant genes lives on to reproduce.

This answer can only gain 2 marks as it does not really discuss the *way* antibiotic resistance in bacteria provides evidence for the theory of evolution. The student needs to use the idea of how bacterial resistance arises and then link that to why it would be evidence for evolution. There needs to be some indication that the student understands that the rapid reproduction cycle of bacteria shows how survival of the fittest leads to the selection of bacteria that are resistant and able to go on to produce offspring that contain their beneficial genes.

Extinction

1. The graph shows the number of species that became extinct between 1800 and 2010.

a Describe the shape of the graph.

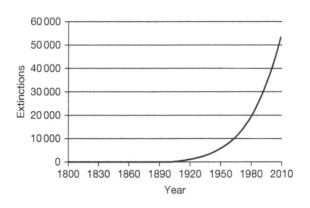

_____

_____ [2 marks]

b Suggest **three** reasons why this may have happened.

1. _____

2. _____

3. _____ [3 marks]

2. When animals die, they usually fall to the ground and decay. In 1977, the body of a baby mammoth was discovered. The baby mammoth died 40 000 years ago and its body froze in ice.

a Explain why the body of the baby mammoth did **not** decay.

_____

_____ [2 marks]

b Mammoths are closely related to modern elephants. Mammoths are now extinct. What does 'extinct' mean?

_____ [1 mark]

Selective breeding

1. Farmers use selective breeding for their animals and crops.

a Give **three** reasons why farmers use selective breeding in animals and plants.

1. _____

2. _____

3. _____ [3 marks]

b Give **two** disadvantages of selective breeding.

1. _____

2. _____ [2 marks]

2. Milk produced by cows has a mean fat content of 3.2%. A farmer has a cow that produces milk with a fat content of 1.6%.

Describe how the farmer could produce more cows that produce low-fat milk.

_____

_____

_____

_____ [3 marks]

3. Dog breeders use selective breeding to ensure that offspring have exaggerated characteristics, for example, shih tzus are bred for their straight fur, curled tail, etc. This selective breeding can also cause breathing problems in shih tzus.

Discuss reasons **for** and **against** a ban on the selective breeding of shih tzus.

_____

_____

_____

_____

_____

_____

_____

_____ [4 marks]

Genetic engineering

1. Genetic engineering involves modifying the genome of an organism by introducing a gene from another organism to give a desired characteristic.

Why are bacterial cells genetically engineered? Tick **one** box.

☐ To be resistant to diseases. ☐ To produce many offspring.

☐ To produce useful substances. ☐ To become resistant to herbicides.

☐ To overcome inherited disorders. [1 mark]

2. Crops with modified genes are called genetically modified (GM) crops.

Why are some people concerned about GM crops? Tick **two** boxes.

☐ Pollen from GM crops may be transported to neighbouring crops.

☐ Not all plant cells take up the foreign gene.

☐ Herbicide-resistant crops may kill non-target organisms.

☐ Some GM crops can resist insect attack. [2 marks]

3. The diagram shows some stages of genetic engineering.

Higher Tier only

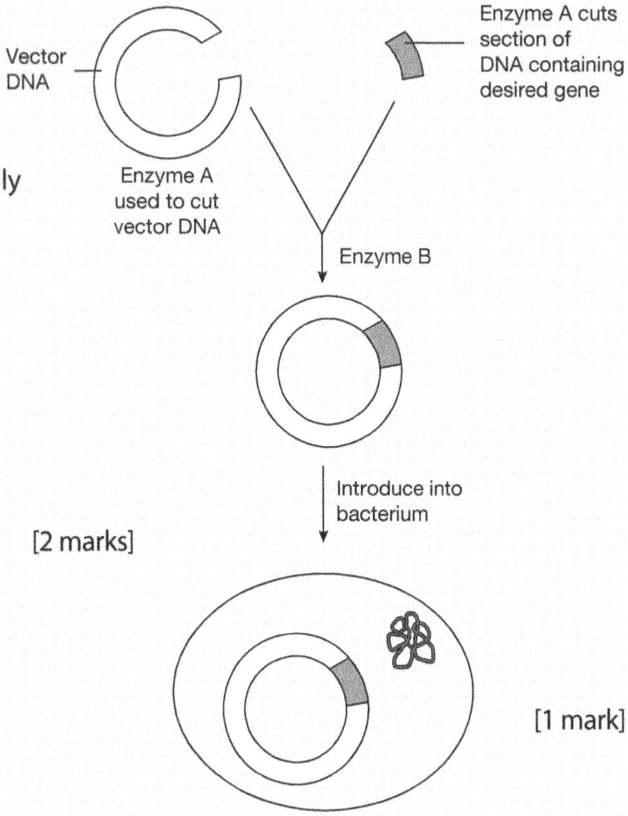

a Genetic engineering uses a vector. Name **two** types of vector commonly used in genetic engineering.

Vector DNA

Enzyme A cuts section of DNA containing desired gene

Enzyme A used to cut vector DNA

Enzyme B

_____

_____

_____

_____ [2 marks]

Introduce into bacterium

b Describe the function of enzyme B.

_____ [1 mark]

c Scientists investigated using genetic engineering to modify cauliflowers to make their leaves more green. The gene for green leaves was inserted into cauliflower cells using a viral vector.

Describe **three** concerns about the use of a viral vector to produce a genetically modified cauliflower.

_____

_____

_____ [3 marks]

4. **a** Describe how the plasmid can be used to genetically modify a bacterial cell to contain a human gene.

Higher Tier only

_____

_____

_____ [3 marks]

b Bacteria can be genetically modified to produce substances that are useful to treat human diseases.

Suggest how **one** named product made by genetically modified (GM) bacteria has an advantage over the same substance made in other ways.

_____

_____ [1 mark]

Classification of living organisms

1. *Agaricus bernardii.* is a mulitcellular organism. It does not contain chlorophyll.

a *Agaricus bernardii* is the binomial name of this organism.

Draw **one** straight line from each part of the binomial name to show which each part of the name refers to.

Binomial name

Classification

| species |
| family |
| phylum |
| genus |
| order |

| Agaricus |
| bernardii |

[1 mark]

b State **two** characteristics of organisms in the plant kingdom.

1. _____

2. _____   [2 marks]

c Vertebrates belong to the animal kingdom. Use words from the box to complete the sentence.

| chordata   crab   fish   mollusca   protoctista |

Vertebrates belong to the phylum _____.

An example of a vertebrate is a _____.   [2 marks]

2. Bacteria are classified as prokaryotes. State **two** characteristics of prokaryotes.

_____

_____ [2 marks]

3. The diagram shows one model of the relationship between some animals.

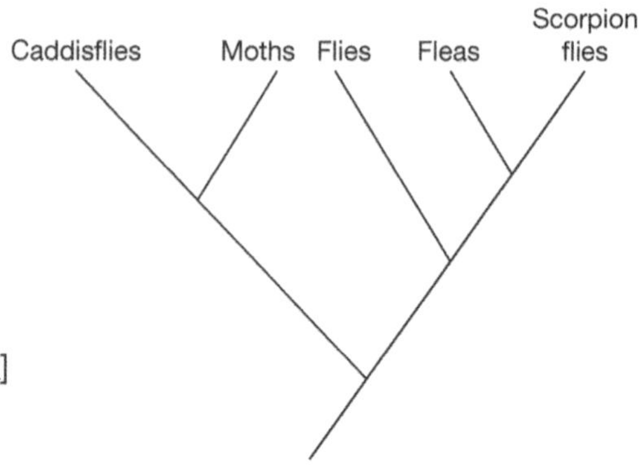

Caddisflies   Moths   Flies   Fleas   Scorpion flies

a Name the type of diagram shown.

_____

[1 mark]

b Name the species most closely related to fleas.

_____ [1 mark]

c Models showing how species are related may change and get updated.

Suggest **one** reason why.

> **Remember**
> The tips of the tree represent the species and the nodes on the tree represent the common ancestors of the species.

_____ [1 mark]

Habitats and ecosystems

1. Use words from the box to complete the sentences.

respiration	germination	pollination	warmth	photosynthesis
	interdependence	shelter	dependency	

The plants and animals in an ecosystem depend on each other in many different ways. Plants

carry out _____, helping to regulate the concentration of oxygen and carbon

dioxide in the atmosphere. Many animals depend on plants for food and _____.

Some plants rely on insects such as bees for _____ and seed dispersal. The way

living organisms interact and rely on each other to survive is called _____.

[4 marks]

2. Which of the following statements **best** describes a community? Tick **one** box.

☐ All the organisms that live in a country.

☐ One species of organism living in an ecosystem.

☐ All the living organisms that interact within the same ecosystem.

☐ The non-living parts of an ecosystem. [1 mark]

3. What is meant by the term 'population'?

_____ [1 mark]

4. The diagram shows the different levels of organisation found within an ecosystem.
Draw **one** line from each label to the matching level on the diagram.

| community |

| ecosystem |

| population |

| individual |

[4 marks]

5. For an ecosystem to be stable and self-supporting, it must have an external source of energy which is usually the Sun or an artificial light source. Explain why this is.

_____

_____

_____

_____ [4 marks]

Food in an ecosystem

1. Add the following labels to the food chain.

| primary consumer | predator | secondary consumer | producer |

strawberry plant → snail → hedgehog → fox

[4 marks]

2. Describe what is meant by the term 'producer'.

_____ [1 mark]

3. What do the arrows in the food chain represent?

_____ [1 mark]

4. A disease kills **most** of the hedgehogs in the food chain.

Describe and explain how this may affect the number of snails and the number of foxes.

_____

_____ [2 marks]

Biotic and abiotic factors

1. Circle **two** biotic factors from the list.

| light | pathogens | temperature | moisture | predators | wind |

[2 marks]

2. Explain what is meant by the term 'abiotic factor'.

_____ [1 mark]

3. List **three** abiotic factors that might affect the distribution of species in a pond.

_____

_____ [3 marks]

4. Which example shows a relationship between both an abiotic and a biotic factor in an ecosystem? Tick **one** box. [1 mark]

☐ Algae being eaten by a fish. ☐ A tree removing a gas from the air.

☐ Water carrying a rock downstream. ☐ A flower providing food for an aphid.

5. Egg wrack is a seaweed that grows on rocky shores. Egg wrack grows on parts of the shore that get covered by water twice a day when the tide comes in.

The graph shows the distribution of egg wrack on a rocky shore. A student was asked to look at the graph and answer the questions below.

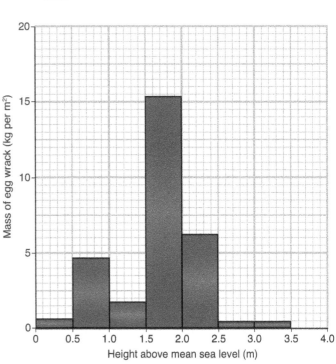

a At what height interval above mean sea level is the largest mass of egg wrack found? [1 mark]

Height interval: <u>1.5 metres</u>

> This would not get the mark. The graph shows the mass of egg wrack in height intervals so the answer is 1.5–2.0 m.

b Describe what happens to the distribution of egg wrack above 2 metres above sea level. Suggest an explanation for the pattern you describe. [2 marks]

The distribution decreases.

This might be because any higher than 2 metres it would be uncovered by sea water for too long.

> This answer is correct. As the command word 'suggest' is used, the student has correctly used 'might' in the response. 'Suggest' questions often have more than one possible answer so words such as might, could, or 'I think that' are appropriate.

6. The graph shows a model predator–prey cycle for foxes and rabbits.

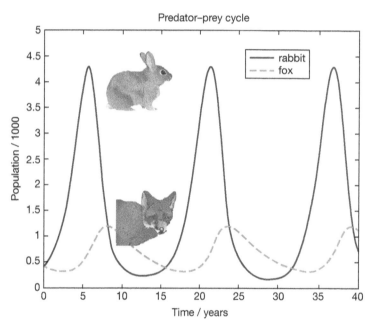

Describe and explain the pattern shown by the graph.

_____

_____

_____

_____

_____

_____

_____

_____ [6 marks]

Adapting for survival

1. Draw **one** line from each animal to show how it is structurally adapted to its environment.

Camel

Polar bear

Toad

Thick fur and blubber

Long sticky tongue

Large surface area to volume ratio

Scales

[3 marks]

2. Describe the **difference** between a structural and a behavioural adaptation.

_____

_____ [2 marks]

3. The plant *Urtica dioica*, commonly known as the stinging nettle, is covered in long thin hollow hairs that contain stinging chemicals. Suggest how this is an adaptation for survival.

_____ [1 mark]

4. Bacteria living in deep sea vents can survive temperatures between 40°C and 80°C. What is the name given to organisms that can survive in extreme environments?

_____ [1 mark]

5. African elephants are the world's largest land mammal. They live in desert environments.

Suggest how the following features help the African elephant to survive there.

Large ears: _____

_____

Tusks: _____

_____ [2 marks]

Measuring population size and species distribution

1.

Practical

A student investigates the effect of trampling on the number of buttercup plants in the school field. She compares an area that is well trampled to an area that is untrampled. Then she samples the buttercups using a 20-metre transect line and a 1 m² quadrat.

Describe, step by step, how the student could do this, taking readings at 5-metre intervals.

_____

_____

_____

_____

_____

_____ [5 marks]

2. The table shows the data collected in the trampled area.

Quadrat sample number	Number of buttercups in trampled area
1	0
2	3
3	0
4	1
5	4

Maths

Make sure you are clear about the difference between mean, median and mode. Think of a way to help you remember the difference. The mean is a bit 'mean' because you have to add up all the numbers and then divide the sum by how many numbers were added. Median sounds like medium or the number in the middle of the data. MOde is the MOST frequent number. They both have 'mo' at the beginning.

Maths **a** Calculate the mean, modal and median number of buttercups for the five quadrat samples.

_____

Mean number of buttercups: _____

Modal number of buttercups: _____

Median number of buttercups: _____   [3 marks]

b The last quadrat placed on the untrampled field was under the shade of a large tree.

Suggest and explain how the tree might affect the number of buttercups.

_____

_____   [2 marks]

Cycling materials

· ·

1. Use the words from the box to complete the sentences below.

carbon   cycled   nitrogen   silver   physical   leached   biotic

Nutrients are chemical elements that all plants and animals require for growth. They include

water, _____ and _____. Each nutrient is _____
from the physical environment into living organisms, and then recycled back to the

_____ environment. This movement of nutrients is a vital function of the
ecology of an area.   [2 marks]

2. The diagram shows the carbon cycle.

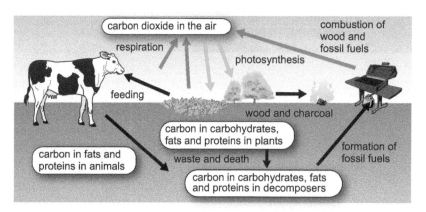

a Give the name of a process that **removes** carbon dioxide from the air.

_____ [1 mark]

b Give the name of **two** processes that release carbon dioxide into the air.

_____ [2 marks]

c Dead organisms are decomposed by microorganisms. Why is this an important part of the carbon cycle?

_____ [1 mark]

d Why does the recycling of carbon and other nutrients take longer in waterlogged soils?

_____

_____ [2 marks]

3.

Maths

In 1995, the amount of carbon released into the atmosphere by burning fuels was 5.3 billion tonnes. In 2015, this had increased to 7.4 billion tonnes.

Calculate the percentage increase in the amount of carbon released. Give your answer to 2 decimal places.

_____

Percentage increase = _____ %   [2 marks]

4. Explain why the water cycle is important to living organisms.

_____

_____

_____ [3 marks]

Effects of human activities

1. Define what is meant by the term 'biodiversity'.

_____

_____ [1 mark]

> **Literacy**
> When trying to remember the meaning of scientific terms, think about their Latin or Greek origins. 'Bio' means living and 'diverse' means having variety.

2. Give **one** reason why biodiversity is beneficial for an ecosystem.

_____ [1 mark]

3. As the human population increases, more land is being used to grow crops. Often one crop is grown over a huge area. What are the negative effects of this change in land use on our ecosystems? Tick **two** boxes.

☐ There is less food available for insect pollinators such as bees.

☐ There is less acid rain.

☐ Habitats are created.

☐ Habitats are destroyed. [2 marks]

4. Peat bogs take thousands of years to form. What is the main reason peat bogs are being destroyed? Tick **one** box.

☐ To produce garden compost. ☐ To clear land for houses.

☐ To produce charcoal. ☐ To prevent the spread of malaria. [1 mark]

5. The rainforests in many countries are under threat from deforestation.

Give **two** reasons why tropical rainforests are being destroyed.

_____

_____ [2 marks]

6. The pie chart below shows what happens to the domestic waste we produce in the UK.

Recycled

Incinerated 3%

Landfill 65%

Maths
It is a straightforward calculation to find out a missing percentage from a pie chart. The 'per cent' means 'out of 100' so remember that all the numbers in a pie chart must add up to 100.

Maths **a** Calculate what percentage of waste gets recycled.

_____

Percentage of waste recycled = _____ % [1 mark]

b Describe **two** ways a landfill site can reduce biodiversity of the area.

_____

_____ [2 marks]

7. Describe and explain the consequences of deforestation on the ecosystem.

_____

_____

_____

_____ [4 marks]

Global warming

1. Which **two** of the following gases contribute to global warming?

| oxygen | carbon dioxide | methane | nitrogen | sulfur dioxide |

[2 marks]

2. The gases that cause global warming occur naturally in the atmosphere, but human activities have increased the level of these gases.

List **two** human activities that have increased the level of these gases.

_____

_____ [2 marks]

3. The graph shows temperature difference compared to 1880 global average temperature between 1880 and 2005.

Describe the pattern shown by the graph. Evaluate to what extent the data provides evidence for global warming. [4 marks]

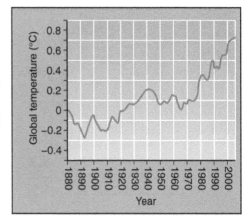

Worked Example

The graph shows that the temperature has fluctuated but overall there is an increase in global temperature.

The graph clearly shows an overall trend of increasing global temperatures. This is strong evidence that global warming is happening as the temperature increase is more than what has been observed in the past.

> The student has correctly described the general pattern shown by the graph and stated the overall trend.

Overall, the evidence for global warming shown by the graph is quite strong.

> The student has remembered to make a concluding remark based on the strength of the evidence.

4. Describe the possible consequences of global warming.

_____

_____

_____

_____ [4 marks]

Maintaining biodiversity

1. Monoculture is the growing of **one type** of crop in a field. How can farms growing monocultures limit the impact of this style of farming on the ecosystem? Tick **one** box.

☐ By removing all hedgerows and trees.

☐ By replanting hedgerows.

☐ By trimming hedgerows.

☐ By adding fertilisers to hedgerows. [1 mark]

2. Farmers are encouraged to reintroduce field margins. Explain how this increases biodiversity.

_____

_____ [2 marks]

3. Many households in the UK now regularly recycle household paper waste such as newspapers.

Explain how this helps protect our ecosystems.

_____

_____

_____ [3 marks]

4. A tree replanting programme claims that replanting trees helps to offset CO_2 emissions from human activities. Evaluate to what extent this claim is true.

_____

_____

_____ [2 marks]

5. Many conservation programmes aim to protect endangered species.

An example is the International Gorilla Conservation Programme in Africa.

Why are conservation programmes such as this beneficial to a country? Tick **two** boxes.

☐ They keep damage to food chains and food webs to a minimum.

☐ They require a lot of funding.

☐ They can attract tourists and therefore benefit the economy.

☐ They prevent endangered species from breeding. [2 marks]

6. The Mangrove Action Project is an organisation that is involved with the conservation and restoration of mangrove forests around the world.

A student's answer to the following question is given below. Suggest **two** difficulties organisations such as the mangrove Action Project may face in trying to conserve and restore the mangrove forests. [2 marks]

Worked Example

The local people might rely on the mangrove forests for food so might not want it to be protected.

Local people or building companies might not know about the importance of the mangrove so might think it doesn't matter if it gets destroyed.

Mangroves are complicated ecosystems that need experts to be involved with restoration and finding experts may be difficult and expensive.

This answer gives three difficulties, which are all correct answers, but the question only asks for **two** so the student would only get 2 marks and has wasted time writing about three issues.

Plant and animal cells (eukaryotic cells)

1. Nucleus – controls the cell's activities; [1 mark] chloroplast – where photosynthesis occurs; [1 mark] cell membrane – controls what enters and leaves the cell; [1 mark] cell wall – for support and protection; [1 mark] cytoplasm – where the cell's activities occur [1 mark]

2. Top: cell membrane; [1 mark] left: nucleus; [1 mark] bottom: cytoplasm. [1 mark]

3. **a** B, C, A [1 mark]

   b As a stain [1 mark] so he can see the organelles. [1 mark]

   c To prevent air bubbles getting trapped under coverslip. [1 mark]

Bacterial cells (prokaryotic cells)

1. Smaller; simple; single; bacteria. [4 marks]

2. A – cell wall; B – cell membrane; C – cytoplasm; D – chromosomal DNA. [4 marks]

3. **a** The cytoplasm. [1 mark]

   b The nucleus. [1 mark]

4. 2/1000 = 0.002 mm. [1 mark]

5. $0.002 = 2 \times 10^{-3}$. [1 mark]

Size of cells and cell parts

1. (Smallest) 1 – ribosome; 2 – mitochondrion; 3 – sperm cell; 4 – egg cell; 5 – nerve cell from giraffe's neck. [5 marks]

2. $\times 400$. [1 mark]

3. 50 mm. [1 mark]

4. **a** $50 \times 1000 = 50\,000$ μm. [1 mark]

   b $50\,000/25 = \times 2000$. [1 mark]

5. ~1 μm (i.e. one-sixth the size of the cell, which is 6 μm). [1 mark]

The electron microscope

1. How much bigger the image of a sample is relative to its actual size. [1 mark]

2. The smallest distance between two points that can still be seen as two points. [1 mark]

3. It has enabled scientists to observe sub-cellular structures in much more detail [1 mark] and to see structures within the cell that are not visible with the light microscope. [1 mark]

4. Advantages:

   Any **two** from: You can obtain good quality images of the internal structure of the cells; they have a better resolution than light microscopes; they have higher magnification than light microscopes. [2 marks]

   Disadvantages:

   Any **two** from: They're very expensive; specialist training is needed to use them; specimen must be dead and prepared in a vacuum. [2 marks]

Cell specialisation and differentiation

1. Unspecialised, differentiate, cilia, specialised. [4 marks]

2. Differentiation [1 mark]

3. They contain many mitochondria. [1 mark]

4. Small size of red blood cells mean they can travel through small capillaries; bi-concave shape gives a large surface area/surface area: volume ratio for absorbing oxygen; contain haemoglobin to carry oxygen; no nucleus so more room for haemoglobin to carry oxygen. [4 marks]

Cell division by mitosis

1. Mitosis produces two new cells that are identical to each other, and to the parent cell. [1 mark]

2. 46 [1 mark]

3. Top row of diagram – the cell grows. The number of sub-cellular structures, e.g. mitochondria, increases. [1 mark] Middle of diagram – the DNA replicates to form two copies of each chromosome. One set of chromosomes is pulled to each end of the cell and the nucleus divides. [1 mark] Bottom row – the cytoplasm and membrane divides and two identical cells are formed. [1 mark]

4. **a** Three. [1 mark]

   b 3×23 hours 18 minutes = 69 hours, 54 minutes; [1 mark] convert to minutes $69 \times 60 = 4140 + 54 = 4194$ minutes. [1 mark]

5. In the nucleus. [1 mark]

6. Number of sub-cellular organelles increases; cell grows. [1 mark]

Stem cells

1. Differentiate, embryos, bone marrow. [3 marks]

2. Any type of human cell. [1 mark]

3. Meristem tissue. [1 mark]

4. It is wrong to destroy life/embryo cannot give permission. [1 mark]

5. Umbilical cord. [1 mark]

6. Nerve cell. [1 mark]

7. Marks in three bands according to level of response.

Level 3 [5–6 marks]: The response gives a detailed and coherent evaluation considering a range of risks and benefits of using stem cells for medical treatments. The response has a conclusion that is consistent with the arguments.

Level 2 [3–4 marks]: The response has an attempt to consider the risks and benefits of using stem cells and a conclusion has been made. The logic may be inconsistent at times, but the response does start to develop a coherent argument.

Level 1 [1–2 marks]: The response lacks structure and contains some unconnected relevant risks and benefits of using stem cells, however the logic is not clear. If there is a conclusion it may not be consistent with the reasoning.

Level 0 [0 marks]: No relevant content.

Points that should be made:

Benefits:

- Many people will be cured of diseases that are currently incurable and so will not suffer from the symptoms e.g.
  - spinal injuries leading to paralysis
  - conditions in which certain body cells degenerate, e.g. Alzheimer's disease, diabetes
  - cancers or following treatments for cancer such as chemotherapy or radiation, e.g. people with leukaemia.
- They enable chemotherapy patients, who have had their bone marrow destroyed, to produce red blood cells.

Risks:

- They may cause unknown long-term side effects.
- There is a chance of rejection if stem cells are not taken from the person that they are being used on.
- Some human embryos will be destroyed during the process.

[6 marks]

Diffusion

1. Molecules from an area of high concentration to an area of low concentration. [1 mark]

2.
cell membrane, which is permeable to oxygen

high concentration of oxygen → low concentration of oxygen

[1 mark]

3. Any **two** from: difference in concentration; temperature; surface area of the membrane. [2 marks]

4. Red blood cells are too big to fit through the capillary walls. [1 mark]

5. If the air is warmer, the molecules of carbon dioxide have more energy [1 mark] and move faster, so the rate of diffusion is faster. [1 mark]

6. Marks in three bands according to level of response.

Level 3 [5–6 marks]: The response gives a clear and detailed explanation of the role of diffusion in gas exchange. The response is well structured.

Level 2 [3–4 marks]: The response gives some explanation of the role of diffusion in gas exchange. The response has some structure and some of the points are linked together.

Level 1 [1–2 marks]: The response gives basic information about the role of diffusion in gas exchange. The answer may lack structure and points are not linked together.

Level 0 [0 marks]: No relevant content.

Points that should be made:

- When you breathe in, the concentration of oxygen molecules in the alveoli is higher than in the capillaries surrounding the alveoli.
- This creates a concentration gradient and therefore…
- Oxygen molecules diffuse through the thin walls of your alveoli and into the capillaries.
- Carbon dioxide concentration in the capillaries surrounding the alveoli is greater than the air breathed in/air in alveoli.
- This creates a concentration gradient.
- Carbon dioxide diffuses from blood into the alveoli and out of the lungs.

[6 marks]

Exchange surfaces in animals

1. Surface area of $3 \times 3 \times 3$ cm cube = 54 cm²; [1 mark] volume = 27 cm³; [1 mark] SA:V = 2:1. [1 mark]

2. Cell A [1 mark] because it has a larger surface area. [1 mark]

3. In a single-celled organism the surface area to volume ratio is big/high [1 mark] so sufficient nutrients such as oxygen can diffuse in. [1 mark] A large multicellular organism, like a human, has a much smaller/lower surface area to volume ratio [1 mark] so it need specialised organs like lungs to exchange materials. [1 mark]

4. It has villi to give a large surface area to maximise absorption. [1 mark] It has thin walls to provide a short diffusion path. [1 mark] It has an efficient blood supply/surrounded by many capillaries. [1 mark]

Osmosis

1. Water, dilute, concentrated. [3 marks]

2. It is a membrane or barrier that allows some molecules or substances to cross, but not others. [1 mark]

3. a To avoid weighing excess water on the surface of the potato. [1 mark]

   b To allow time for osmosis to take place. [1 mark]

   c Both axes correctly labelled; points plotted correctly; correct line of best fit. [3 marks]

d Accept between 11.5% and 13.5%. [1 mark]

e Yes, because at this concentration there was no change in mass [1 mark] because concentration in boiling tube and potato cells was the same so no osmosis took place. [1 mark]

Active transport

1. Lower, higher, against, energy. [4 marks]

2. The soil contains a very dilute concentration of mineral ions compared to the root hair, so the ions need to be moved against a concentration gradient requiring active transport. [2 marks]

3. Active transport works against a concentration gradient whereas diffusion works along a concentration gradient. Active transport requires energy from respiration whereas diffusion does not require energy. Both are forms of transport that move substances. [4 marks]

Section 2: Organisation

Digestive system

1.

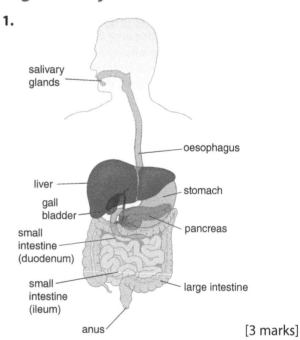

salivary glands

oesophagus

liver

stomach

gall bladder

small intestine (duodenum)

pancreas

small intestine (ileum)

large intestine

anus

[3 marks]

2. An organ system in which several organs work together to break down and absorb food. [1 mark]

3. To break down food so it can be absorbed and used by the body. [1 mark]

4. Physical digestion is the physical breaking up of food into smaller pieces; chemical digestion is the breakdown of food into small molecules by enzymes so they can be absorbed by the blood. [2 marks]

Digestive enzymes

1. Small, bloodstream, proteins, respiration. [4 marks]

2. Top row: Amylase is produced in the salivary glands [1 mark] and breaks down starch into glucose. [1 mark]

Middle row: protease. [1 mark]

Bottom row: lipids. [1 mark]

3. Benedict's test – carbohydrates – sugars. [1 mark]
Biuret test – proteins. [1 mark]

Iodine test – carbohydrates – starch. [1 mark]

4. a Grinding – to increase the surface area. [1 mark] Adding distilled water – to make a solution. [1 mark]

b Biuret test: Add a few drops of biuret solution into the milk solution; the solution will turn from blue to purple/violet colour if protein is present. [2 marks]

Factors affecting enzymes

1. a The iodine will no longer turn blue-black. [1 mark]

b The temperature of the water bath is much easier to control. [1 mark]

c When the sample is tested at regular intervals to find out if the starch has been broken down. [1 mark]

d 40 degrees [1 mark]

e The rate of reaction reaches a maximum [1 mark] because many fast-moving substrate molecules enter and fit easily into the active site. [1 mark]

Heart and blood vessels

1. a Clockwise from top left: pulmonary artery, aorta, pulmonary vein, vena cava. [4 marks]

b Arrow into pulmonary vein. [1 mark]

c Arrow out of aorta. [1 mark]

2. A pacemaker is a group of cells in the right atrium of the heart. [1 mark]

Artificial pacemakers can be used to correct irregularities in the heart rate. [1 mark]

3. The walls are thin [1 mark] and permeable. [1 mark]

4. Marks in three bands according to level of response.

Level 3 [5–6 marks]: The response gives a clear, logical explanation of a range of both similarities and differences between arteries and veins.
Level 2 [3–4 marks]: The response has some key ideas about at least one of the similarities and at least one of the differences between arteries and veins. The ideas may not be expressed in a logical manner.
Level 1 [1–2 marks]: The response lacks structure and contains some unconnected relevant key ideas of at least one of the similarities or at least one of the differences between arteries and veins.
Level 0 [0 marks]: No relevant content.

Points that should be made:

Similarities:

- They both have walls made of three layers.
- They both have a lumen.
- They both transport blood and dissolved substances around the body.

Differences:

- Arteries carry blood from the heart to the rest of the body, but veins carry blood from the body to the heart.
- Most arteries carry oxygenated blood and most veins carry deoxygenated blood, except the pulmonary artery, which carries deoxygenated blood from the heart to the lungs, and the pulmonary vein, which carries oxygenated blood from the lungs to the heart.
- Arteries have a thicker (elastic) muscle wall/veins have a much thinner wall.
- Arteries carry blood with a higher pressure and a pulse/veins carry blood at much lower pressures with no pulse.
- Arteries have a narrower lumen/veins have a wider lumen.
- Veins have valves and arteries do not.

[6 marks]

Blood

1. Red blood cells – transport oxygen around the body. [1 mark]

   White blood cells – protect the body from infection. [1 mark]

   Platelets [1 mark] – help the clotting process at wound sites.

2. Any **three** from: hormones, antibodies, nutrients (glucose, amino acids, minerals, vitamins), waste substances (carbon dioxide, urea). [3 marks]

3. If they are unwell/fighting an infection/cancer. [1 mark]

4. It gives it a large surface area [1 mark] to allow greater absorption of oxygen. [1 mark]

5. **a** 3575 cm^3 (55/100 × 6 500 cm^3). [1 mark]

   b 5.0×10^6. [1 mark]

Heart–lungs system

1. Trachea, bronchus, bronchiole, alveoli. [1 mark]

2. The blood flows in two circuits [1 mark] – one from heart to lungs and one from heart to rest of body. [1 mark]

3. Any **four** points from: the spherical shape gives a large surface area to volume ratio, which enables efficient diffusion of gases; [1 mark] the wall of alveoli is very thin so gases do not have far to diffuse; [1 mark] each alveolus is surrounded by a network of capillaries/good blood supply so oxygen is constantly moved into the blood and carbon dioxide moved into the lungs; [1 mark] this means gas exchange happens at the steepest concentration gradient possible; [1 mark] alveoli surfaces are moist so gases dissolve, allowing efficient diffusion. [1 mark]

Coronary heart disease

1. This disease is not passed from one person to another/is not infectious. [1 mark]

2. C, A, D, B [4 marks]

3. It may result in backflow [1 mark] so blood is pumped much less efficiently; [1 mark] **or** oxygenated and deoxygenated blood may mix [1 mark] so oxygen is not pumped as efficiently. [1 mark]

4. Advantage: no rejection. [1 mark]

   Disadvantage: patient needs anti-clotting drugs for rest of life or artificial valve can damage red blood cells. [1 mark]

5. Stent: opens up an artery to allow blood to supply the heart muscle with glucose and oxygen for respiration. [1 mark]

   Statins: stop the liver producing as much cholesterol, so less narrowing of arteries. [1 mark]

6. Any **two** from: Change diet so less cholesterol/fat; exercise more; stop smoking; stop drinking alcohol. [2 marks]

7. The answer must have the benefit [1 mark], three risks [3 marks] and a judgement [1 mark].

   The benefits are that with a heart transplant the person will live longer and have a better quality of life because they will have more energy and strength with a new heart.

   Any **three** risks from:

   The person has to have surgery, which is dangerous if it goes wrong.

   The person could bleed to death or get an infection from the operation.

   The person would also have to take anti-rejection drugs which might have side effects.

   The person might have to wait a long time for a donor heart because there is a shortage of heart donors, so they might die waiting for a new heart.

   A judgement such as:

   Overall, the benefits outweigh the risks because the person will be alive longer with a heart transplant. The person would probably die sooner if they did not have the transplant.

Risk factors for non-infectious diseases

1. Tobacco use, physical inactivity, unhealthy diet and the harmful use of alcohol. [1 mark]

2. Any **three** from: gender, age, diet, whether the person smokes or not, alcohol consumption, genetics. [3 marks]

3. Many diseases are caused by an interaction of a number of factors. [1 mark] It is therefore difficult to collect data that prove causal mechanism. [1 mark]

4. **a** Accept 2500–3000. [1 mark]

   b More older people suffer coronary heart disease than younger people. [1 mark]

   c Women live longer. [1 mark]

Cancer

1. Strict, mutation, tumour, lifestyle. [4 marks]
2. Any **three** from: smoking, drinking alcohol, being overweight, an unhealthy diet, lack of exercise, overexposure to ionising radiation, viruses. [3 marks]
3. Viruses living in cells can be the trigger for certain cancers. [1 mark] As a tumour grows, cancer cells can detach and spread to other parts of the body. [1 mark]
4. Chemical or other agent that causes cancer. [1 mark]
5. Any **two** from: A benign tumour is slow-growing, a malignant tumour is faster; a benign tumour often has a clear border, a malignant does not necessarily; a benign tumour is not cancerous, a malignant tumour is; a benign tumour rarely spreads, a malignant tumour spreads to other body tissues easily. [2 marks]

Leaves as plant organs

1. Clockwise from the top of the diagram: palisade cell, spongy mesophyll cell, stoma, guard cell. [4 marks]
2. Because it is a group of tissues performing a function (photosynthesis). [1 mark]
3. Xylem [1 mark] and phloem. [1 mark]
4. The shoots of a plant. [1 mark]
5. Broad leaves: give large surface to absorb maximum amount of light. [1 mark]

   Palisade cells: contain many chloroplasts/ arranged end-on to absorb maximum amount of light. [1 mark]

   Thin and transparent upper epidermis: allows maximum light to pass through to palisade cells. [1 mark]

Transpiration

1. Light intensity, temperature, wind. [3 marks]
2. The movement of water through the plant and leaves. [1 mark]
3. $5 \times 60 = 300$ seconds; [1 mark] $9/300 = 0.03$. [1 mark]
4. **a** Any **two** from: temperature, air movement/ wind, light, humidity. [2 marks]

   b Plant A lost more water ($252 - 239 = 13$ g) than plant B ($137 - 129 = 8$ g). [1 mark] The rate of transpiration was higher in plant A/ the plant with broad flat leaves. [1 mark] This is because a broad flat leaf loses more water than a needle due to larger surface area/ more stomata for water to diffuse from. [1 mark]

Translocation

1. The movement of dissolved sugars around the plant. [1 mark]
2. Some is stored and some is used for respiration. [1 mark]
3. They have a large surface area to maximise the absorption of water. [1 mark]
4. Marks in three bands according to level of response.

Level 3 [5–6 marks]: The response gives a clear and detailed comparison of the structures and functions of xylem and phloem. The response is well structured.
Level 2 [3–4 marks]: The response gives some comparisons of the structures and functions of xylem and phloem. The response has some structure.
Level 1 [1–2 marks]: The response gives basic information about the structure and function of xylem and phloem but no or limited comparisons. The answer lacks structure.
Level 0 [0 marks]: No relevant content.

Points that should be made:

Structure:

- Both xylem and phloem are found in vascular bundles.
- Phloem cells are elongated, thin-walled, living cells that form tubes.
- Xylem tubes are made from dead cells that are strengthened by lignin and form hollow tubes.
- Phloem contains sieve plates.
- Xylem does not contain sieve plates.

Function:

- Phloem transports dissolved sugars from the leaves to the rest of the plant.
- Transport of substances in the phloem can be in any direction.
- Xylem transports water and minerals from the roots to the stem and leaves.

- Transport of water and minerals is in one direction.
- Phloem is transport mechanism for translocation.
- Xylem is transport mechanism for transpiration.

Section 3: Infection and response

Microorganisms and disease

1. Communicable: Any **one** from: measles, mumps, rubella, colds, flu, impetigo, any other infectious disease. [1 mark]

   Non-communicable: Any **one** from: diabetes, coronary heart disease, stroke, cancer. [1 mark]

2. Viruses – flu; protists – malaria; fungi – Athlete's foot; bacteria – food poisoning. [4 marks]

3. Yes: many women with cervical cancer have HPV16 (18, 31). [2 marks] **or** No: few women with cervical cancer have HPV 6 or 11. [2 marks] The other marks for any **two** points from: HPV does not mean causation because it may be caused by another factor/due to coincidence; did not study HPV in healthy women; having cancer may cause susceptibility to HPV; does not add up to 100%; not all women with cancer have HPV.

4. Fungi. [1 mark]

5. **a** Cells cannot photosynthesise/make food. [1 mark]

   b Any **four** from: fungus produces spores; spores released in wet, humid conditions/when it rains or the plants are watered; spores dispersed by wind; optimal temperature for fungal growth (24°C)/wet, hot conditions needed for spores to germinate; symptoms start to appear on leaves 3–10 days after infection; spores produced throughout growing season; spores survive on dropped leaves/in soil. [4 marks]

   c The disease is caused by fungus that produces spores, so it is really important to make sure that the affected leaves and stems are removed immediately and burned. If the infected plant is allowed to remain untreated, the spores can be spread by rain or wind. In addition, fungicides can help kill the fungus. Also, infected parts of the plant should not be composted as spores can survive and re-infect other rose plants. [4 marks]

Viral diseases

1. Viruses live inside cells. [1 mark]

2. Sexual contact; exchange of bodily fluids/blood; sharing needles. [2 marks]

3. Small amounts of dead or weakened virus that causes measles. [1 mark]

Bacterial diseases

1. Bacteria can infect plants and animals. [1 mark]

2. **a** Eating infected food [1 mark] and not washing hands before preparing food. [1 mark]

   b The bacteria need time to multiply [1 mark] and toxins build up [1 mark] causing the symptoms to develop. [1 mark]

3. Pain when urinating [1 mark]; yellow or green discharge from penis or vagina. [1 mark]

4. Antibiotics; [1 mark] condoms/abstinence. [1 mark]

Malaria

1. Protists. [1 mark]

2. **a** Liver. [1 mark]

   b Vector. [1 mark]

3. Draining stagnant water pools: Mosquitoes lay eggs on still water; [1 mark] removing water will prevent breeding mosquitoes. [1 mark]

   Using mosquito nets: Mosquitoes bite humans for a blood meal; [1 mark] if an infected person is bitten the disease is transmitted to a non-infected person at the next meal. [1 mark]

Human defence systems

1. White blood cells – produce antimicrobial substances; Stomach acid – kills the majority of pathogens that enter via the mouth; Platelets – form scabs which seal the wound. [3 marks]

2. Diagram showing the white blood cell completely surrounding the pathogen. [2 marks]

3. Hairs in the nose trap larger microbes/dust particles; [1 mark] goblet cells produce mucus; [1 mark] sticky mucus traps microbes; [1 mark] ciliated epithelium/cilia beat to waft mucus away. [1 mark]

Vaccination

1. Lives inside cells; inactive; antibodies. [3 marks]

2. a 80 per 10 000. [1 mark]

   b It increased. [1 mark]

   c The number of children diagnosed with autism continued to rise after the MMR vaccination programme was stopped. [1 mark]

3. (Flu) viruses frequently mutate into new strains; [1 mark] antigens on pathogen are changed; [1 mark] memory lymphocyte not able to recognise new antigen; [1 mark] new vaccines need to be made for flu/understanding that only some viruses mutate. [1 mark]

Antibiotics and painkillers

1. Antibiotics – kill bacteria by interfering with the process that makes bacterial cell walls; [1 mark] Painkillers – relieve symptoms of infection. [1 mark]

2. a Bacteria mutate/there is variation in bacteria; [1 mark] leading to bacteria that survive the antibiotic; [1 mark] these bacteria go on to breed. [1 mark]

   b Doctors should avoid over-prescribing antibiotics [1 mark], patients should take full course of antibiotics. [1 mark]

3. Antibiotics can only kill bacteria; [1 mark] flu is caused by a virus not bacteria. [1 mark]

4. Cannot kill viral/fungal pathogens/protists; [1 mark] specific antibiotics for specific bacteria; [1 mark] emergence of resistant bacterial strains. [1 mark]

Making and testing new drugs

1. To check they work effectively; to check the right dose; to make sure they are safe to use. [2 marks]

2. Cells or tissues. [1 mark]

3. Heart drug digitalis – foxgloves; painkiller aspirin – willow trees; anti-malarial quinine – tree bark; antibiotic penicillin – mould. [2 marks]

Section 4: Photosynthesis and respiration reactions

Photosynthesis reaction

1. Leaves, light, chloroplasts. [3 marks]

2. Carbon dioxide, oxygen. [2 marks]

3. Endothermic. [1 mark]

4. Respiration. [1 mark]

5. Photosynthesis is the reaction that produces glucose. Glucose is the building block for all the tissues that make up plants/producers. Animals rely on plants/producers or other animals that eat plants for energy. [2 marks]

Rate of photosynthesis

1. Low light intensity; low carbon dioxide concentration; low temperatures (also accept low amount of chlorophyll). [3 marks]

2. a Marks in three bands according to level of response.

Level 3 [5–6 marks]: The response gives a clear description of a method with apparatus that would produce valid results. A description of how the rate of photosynthesis is measured is included.
Level 2 [3–4 marks]: A method involving pondweed and varying light intensity is given. A description of what is measured, or at least one control variable is included.
Level 1 [1–2 marks]: The response includes simple statements relating to relevant apparatus or the method involving pondweed and light.
Level 0 [0 marks]: No relevant content.
Points that should be made: • Description of how the apparatus would be used. • Use of ruler to measure distance of light from beaker/pondweed. • Reference to varying distance of light from pondweed. • Accept alternative methods to alter light intensity.

- Measure number of bubbles/volume of gas produced.
- Same length of time.
- Reference to control of temperature.
- Reference to control of carbon dioxide in water.
- Do repeats and calculate a mean.

b 42.3, 31.3, 22.7. [3 marks]

c 1. Bubbles may be of different sizes. [1 mark]
2. It may be difficult to count bubbles if there are lots of them. [1 mark]

d As the light intensity decreases, the rate of photosynthesis also decreases. [1 mark]

3. 0.0025. [1 mark]

4. Quarter the intensity. [1 mark]

Limiting factors

1. The plants/algae in the pond will have been photosynthesising all day/since sunrise. Photosynthesis produces oxygen so oxygen concentrations will have increased throughout the day. [2 marks]

2. Increased carbon dioxide. [1 mark] Increased temperature. [1 mark]

3. a Between A and B the rate of photosynthesis increases linearly with carbon dioxide concentration, because carbon dioxide is needed for photosynthesis, so the more there is, the faster the rate. Between B and C the rate does not increase because other factors such as light intensity become limiting. [4 marks]

b Yes, [1 mark] because the concentration of carbon dioxide is the limiting factor up until about a concentration of 15% of carbon dioxide in the air. [1 mark]

Uses of glucose from photosynthesis

1. Starch, storage, growth, respiration. [4 marks]

2. Plant cells respire all the time, but some also carry out photosynthesis when light is available. [1 mark] Only some plant cells can carry out photosynthesis; some, such as root hair cells, do not. [1 mark]

3. From the starch stored in the potato tuber. [1 mark]

4. The part of the potato plant above the soil can photosynthesise and therefore produce glucose for energy. [1 mark]

5. Because starch is insoluble and therefore better for storage. Glucose is soluble. [1 mark]

6. Because nitrate ions are needed to form amino acids, which are needed to make proteins. [1 mark] Proteins are required for healthy growth. [1 mark]

Cell respiration

1. Glucose, aerobically, anaerobically, mitochondria. [4 marks]

2. Respiration releases energy from glucose. [1 mark]

3. Oxygen; water. [2 marks]

4. Any **two** from: movement; keeping warm; chemical reactions to build larger molecules; active transport; cell division. [2 marks]

5. Because energy is transferred to the surroundings during the reaction. [1 mark]

6. Amino acids join together to make proteins; [1 mark] glucose molecules join together to make glycogen. [1 mark]

7. The carbohydrate provides the glucose needed for respiration. [1 mark] The runner's muscle cells will use lots of energy to contract during the race, so a good supply of glucose will enable the muscles to respire at a faster rate. [1 mark]

Anaerobic respiration

1. Without oxygen. [1 mark]

2. Lactic acid. [1 mark]

3. Normally we have enough oxygen for aerobic respiration to take place. [1 mark] Aerobic respiration is preferable as it produces more energy/is more efficient. [1 mark]

4. To begin with, his muscle cells respire aerobically because they have enough oxygen. [1 mark] After several minutes of sprinting his circulatory system is not able to provide enough oxygen for aerobic respiration in his muscle cells. [1 mark] The cells run out of oxygen so they switch to anaerobic respiration. [1 mark]

5. Ethanol + carbon dioxide. [2 marks]

6. Manufacture of bread and alcohol. [2 marks]

7. Yeast cells produce ethanol, whereas muscles produce lactic acid. [1 mark] Yeast produces carbon dioxide, muscle cells do not. [1 mark] Both produce little energy compared to aerobic respiration. [1 mark]

Response to exercise

1. Breathing volume, oxygenated, glucose, carbon dioxide. [4 marks]
2. They work less efficiently. [1 mark]
3. Because the body needs to get rid of the lactic acid that has built up, which needs extra oxygen/oxygen debt. [1 mark] By continuing to breathe heavily, a person can take in the extra oxygen needed to react with the accumulated lactic acid and remove it from the cells. [1 mark]
4. a B. [1 mark]
   b A. [1 mark]
   c C. [1 mark]
   d Because the oxygen debt is the amount of oxygen needed during exercise minus the amount of oxygen absorbed. [1 mark]

Section 5: Automatic control systems in the body

Homeostasis

1. Blood glucose levels and body temperature. [2 marks]
2. Brain/hypothalamus. [1 mark]
3. It keeps the body at an optimum temperature. [1 mark]
4. False, true, true, false, true. [5 marks]
5. Lungs – carbon dioxide; [1 mark] kidneys – urea/any other substance in urine (not glucose); [1 mark] skin – water/salt. [1 mark]

The nervous system and reflexes

1. Receptors, sensory neurone, relay, effector. [4 marks]
2. a 1.26/90 = 0.014 seconds [1 mark]
   b Additional time taken for nerve impulse to travel through synapses/between nerve cells. [1 mark]
3. Any **two** from: drop the ruler from the same height each time; let the ruler drop without using any force; same type/weight of ruler; thumb should be same distance from the ruler each time at the start; use the same hand

to catch the ruler each time; carry out the experiment with the lower arm resting in the same way on the table. [2 marks]
4. No indication beforehand when the colour will change/you might be able to tell when the person is about to drop the ruler; measurement of time is more precise (than reading from a ruler)/resolution (of computer timer) is higher. [2 marks]

Hormones and the endocrine system

1. a They are carried in the blood. [1 mark]
   b Glands. [1 mark]
2. **A** = Pituitary (gland); **B** = adrenal (gland). [2 marks]
3. a Pituitary [1 mark]
   b Regulates secretion of other glands. [1 mark]
4. a Pituitary (gland); thyroxine (accept thyroid hormone); adrenal (gland); oestrogen. [4 marks]
   b The pituitary gland produces thyroid-stimulating hormone which acts on the thyroid gland causing it to release thyroxine. It also produces the hormone FSH that makes the ovaries release oestrogen. [6 marks]

Controlling blood glucose

1. a Patient A. [1 mark]
   b Lethargy/thirst. [1 mark]
   c To convert glucose/remove glucose from the blood; their pancreas is unable to produce insulin. [2 marks]
2. Too much glucose removed from blood; fainting/coma. [2 marks]
3. b Adults, normally over the age of 40 [1 mark]; d linked to poor diet/obesity [1 mark]; e pancreas does not make enough insulin causing high levels of glucose in the blood [1 mark]; h Exercise and controlled diet (especially carbohydrates) [1 mark]
4. Glucagon is secreted; converts stored glycogen into glucose; glucose released into the bloodstream. [3 marks]

Hormones in human reproduction

1. Ovaries, testosterone. [2 marks]
2. Ovary, uterus, fertility drugs. [3 marks]

3. Causes an egg to mature in the ovaries [1 mark]; stimulates ovaries to produce oestrogen. [1 mark]

Hormones interacting in human reproduction

1. a LH. [1 mark]

b LH production is stimulated. [1 mark]

c Any **two** from: placenta development; maintain blood supply; supply nutrients/ remove waste products. [2 marks]

2. Marks in three bands according to level of response.

Level 3 [5–6 marks]: The response gives a clear, logical explanation containing accurate ideas about oestrogen and progesterone presented in the correct order in the cycle.
Level 2 [3–4 marks]: The response has some key ideas about oestrogen and progesterone presented to form a partial explanation.
Level 1 [1–2 marks]: The response has fragmented ideas about one or both hormones, that are not necessarily in the order of the cycle. Some may be relevant, but there are insufficient links to form an explanation.
Level 0 [0 marks]: No relevant content.
Points that should be made:
• Oestrogen levels are low in the early part of the cycle.   • Oestrogen levels increase prior to ovulation.   • After ovulation, oestrogen levels drop.   • From day 1 to day 14 progesterone levels are low.   • Progesterone levels rise after ovulation.   • Hormone levels drop if the ovum is not fertilised.   • Menstruation occurs between day 1 and days 5/6 (this is when the lining of the uterus is shed).   • The lining of the uterus is then built back up (to prepare for fertilised ovum).   • At day 14 ovulation occurs.

[6 marks]

Contraception

1. a Female sterilisation. [1 mark]

b Diaphragm. [1 mark]

c Plastic IUD advantages: It can decrease period cramping, side effects are short term and not as severe as those for the copper IUD. [1 mark]

Plastic IUD disadvantages: It is not effective for as long as the copper IUD so would need to be replaced more often. It does have side effects like cramping and backache after insertion and it can cause bleeding between periods for the first 3 months. [1 mark]

Copper IUD advantages: It is effective for longer than plastic IUD and doesn't need replacing so often. It can be used as emergency contraception. [1 mark]

Copper IUD disadvantages: it can cause heavy and painful periods and bleeding between periods. [1 mark]

Judgment made based on evidence. [1 mark]

2. Birth control pills are 99% effective in preventing pregnancy; The hormones in the pills give protection against some women's diseases; The woman's monthly periods become more regular. [3 marks]

Using hormones to treat infertility

1. FSH; LH. [2 marks]

2. Any **three** from: may increase chance of getting a sexually transmitted disease; may cause side-effects on female body; prolonged use may prevent later ovulation; may cause multiple births. [3 marks]

3. IVF advantages: A higher success rate. [1 mark] IVF disadvantages: Fertility drugs injected carry risk of ovarian hyperstimulation, more expensive/double the cost of IVM. [1 mark] IVM advantages: No fertility drugs injected so no risk of ovarian hyperstimulation, cheaper than IVF/half the cost of IVF. [1 mark] IVM disadvantages: Much lower success rate/success rate half that of IVF. [1 mark] Appropriate judgement based on advantages and disadvantages stated in answer. [1 mark]

Negative feedback

1. a Vasodilation; sweating. [2 marks]

b Any **three** from: when temperature falls; brain detects this; sends message to muscles and glands/increase in temperature; when body temperature becomes high; brain detects and stops sending messages; negative feedback because when desired effect is reached system is turned off. [3 marks]

2. The increase in blood glucose level after the meal is detected by receptors [1 mark] which stimulate the release of insulin from the pancreas/endocrine gland. [1 mark] The insulin simulates the conversion of glucose to glycogen/acts to lower the blood glucose level back to its normal level. [1 mark] The lowering of the blood glucose level inhibits the production of insulin in a negative feedback loop. [1 mark]

Section 6: Inheritance, variation and evolution

Sexual reproduction and fertilisation

1. a Sexual reproduction. [1 mark]
   b The features are inherited/controlled by genes. [1 mark] Genes/DNA/genetic material from both the lion and the tiger are passed on to the gamete [1 mark] and therefore the liger will have features of both. [1 mark]

2. a Sexual/sex. [1 mark]
   b Either sperm or egg cell circled. [1 mark]
   c Mitosis: Y [1 mark]
     Meiosis: X [1 mark]
   d Any **two** from: genes/genetic information/chromosomes from two parents; alleles may be different; environmental effect/there may have been mutation. [2 marks]

Asexual reproduction

1. Asexual; gamete; clone. [3 marks]
2. a Pollen and egg cells. [1 mark]
   b Sperm and egg cells. [1 mark]
3. Potatoes produced by asexual reproduction will be identical to their parent [1 mark], but potatoes produced by sexual reproduction will be different to their parents. [1 mark]

Cell division by meiosis

1. Ovaries, testes. [2 marks]
2. 14 [1 mark]
3. a One solid and one dashed chromosome in each cell; different length chromosomes in each cell. [2 marks]
   b Nucleus. [1 mark]
   c Ovaries. [1 mark]
   d Mitosis produces cells that are diploid/with a full set of chromosomes; whereas meiosis produces cells that are haploid/half the number of chromosomes. [2 marks]

DNA, genes and the genome

1. Sex, genes, chromosomes, nucleus. [4 marks]
2. Chromosomes. [1 mark]
3. Any **two** from: better preventative medicine; identify targets of drugs more effectively/tailor healthcare/personalised medicine; search for genes linked to different types of disease; tracing human migration patterns from the past. [2 marks]
4. a A, T, C and G. [1 mark]
   b A–T and C–G. [1 mark]
5. Any **three** from: some people may be pressured to not have children or to terminate pregnancies; there may be increased pressure for germ line therapy to prevent children inheriting genetic conditions; it may result in discrimination e.g. in the workplace; insurance companies may charge increased premiums to people likely to develop a condition; some parent(s) may want 'designer babies' with selection of gender or specific fashionable characteristics; knowing that you may develop a condition etc. can lead to psychological stress; not everyone will want to know if they are likely to get a condition; it may be perceived as an invasion of human rights/personal freedom (intrusion, infringement of civil liberties); who decides who should have genetic tests/who decides who has access to or should have potentially expensive treatment/data protection issues. [3 marks]

Inherited characteristics

1. Genotype; only expressed; homozygous. [3 marks]

2. Genotype of mouse parent A is Bb [1 mark]; alleles in gametes B b b b [1 mark]; genotypes of offspring Bb Bb bb bb [1 mark]; phenotypes of offspring black fur, black fur, brown fur, brown fur. [1 mark]

3. a Punnett square drawn with alleles of parents Tt down side and across top [1 mark], correct genotypes of offspring TT Tt Tt tt. [1 mark] [2 marks]

b 3:1 [1 mark]

c i The genotype is the alleles the organism has [1 mark] whereas the phenotype is the actual characteristics an organism has. [1 mark]

ii Alleles are different forms of a gene controlling a characteristic and occupying the same site on homologous chromosomes (e.g. B or b); genes are the units of DNA/sites on chromosomes carrying the information that determines characteristics (e.g. bB). [2 marks]

iii Homozygous: BB/bb/possessing a pair of identical alleles for a character/true breeding. Heterozygous: Bb/carrying a pair of contrasting/different alleles for a characteristic. [2 marks]

Inherited disorders

1. a Father: Xh Y; mother: XH XH. [2 marks]

b Phenotype: carrier; Genotype: XH Xh. [2 marks]

c Receives normal XH from mother; Y from father. [2 marks]

d One quarter/one out of four/25%/1:4. [1 mark]

e Female with haemophilia must get Xh from both parents/father must have haemophilia and mother a carrier. [2 marks]

2. a Punnett square completed with genotypes of all offspring FF Ff Ff ff [1 mark], circle around ff. [1 mark]

b 0.25/25%/1:4/1 out of 4. [1 mark]

c Heterozygous. [1 mark]

3. a Yes, because Alex's mother is homozygous recessive/pp [1 mark] so Alex's father must have at least one dominant allele. [1 mark]

b Correct parental alleles [2 marks] correct genotypes of offspring [1 mark] correct probability. [1 mark]

		Father	
		P	p
Mother	p	Pp	pp
	p	Pp	pp

Probability = 50% or ½ or 1:1

Sex chromosomes

1. a Either of the single X boxes under 'mother'. [1 mark] **b** XY. [1 mark]

2. 50% [1 mark]

3. a Male alleles XY, gametes alleles X Y, possible offspring alleles XX, XY, XX and XY. [3 marks]

b 1:1 or 50% or ½ or 0.5 or 1 in 2 or 1 out of 2 or 50:50 (do not accept 50/50). [1 mark]

Variation

1. a Differences between members of the same species/organisms of the same kind/type/Bizzy Lizzies. [1 mark]

b Plants nearest the fence had less sunlight/less warmth from the sun/more shade or converse answer. [2 marks]

c i Genetic and environmental [1 mark]

d They had the same genes/DNA (or similar statement)/they are a clone. [1 mark]

2. Bacteria mutate or idea that there is variation in bacteria; leading to bacteria/resistant cells that survive antibiotics; these bacteria (resistant cells) go on to breed; do not allow bacteria to get used to antibiotics or idea that antibiotics change the bacteria or bacteria become immune or references to adaptation or evolution. [3 marks]

3. a Genetics/inherited/from parent/genes/mutation; environment/surroundings. [2 marks]

b Any **two** from: variation caused by genetic or environmental factors; mutations causing variation in the population; large gene pool for this population, which means a large number of combinations of genes are possible. [2 marks]

Evolution by natural selection

1. Natural selection; simple life forms. [2 marks]
2. **a** Natural. [1 mark]
   b Three billion. [1 mark]
3. Any **four** from: there is variation (of characteristics) within a species; (caused by) different genes; some individuals are better adapted to their environment than other individuals; some phenotypes are better suited to environment; these are more likely to survive and breed; these variations/genes (are more likely to be) passed on to next generation. [4 marks]
4. Each population will adapt to the different conditions in their environment to survive in it [1 mark]; mutations can also occur in the population. [1 mark] This means that each population cannot then interbreed to produce fertile offspring with the other population [1 mark] so a new species is developed. [1 mark]

Fossil evidence for evolution

1. **a** Fish. [1 mark]
   b Birds. [1 mark]
   c The theory of evolution. [1 mark]
2. **a** D [1 mark]
   b It has the right conditions for fossilisation such as no oxygen [1 mark] right amount of heat/pressure to turn sediment into rock. [1 mark]

Other evidence for evolution

1. **a** *Fusarium*. [1 mark]
   b Any **three** from: banana plants genetically identical/clones; same susceptibility/lack of resistance to Panama disease; no/little mutation; close planting enabled easy spread. [3 marks]
   c Any **four** from: variation between individuals; chance/random mutation; some fungi with the variation not affected

by fungicide; fungi carrying resistant genes not killed/survive; mutation/characteristic ability to survive pesticide passed on to offspring. [4 marks]

2. Marks in three bands according to level of response.

Level 3 [5–6 marks]: The response gives a clear, logical explanation containing accurate ideas about how antibiotic resistant bacteria arise and explains how this is evidence for the theory of evolution.
Level 2 [3–4 marks]: The response has some key ideas about how antibiotic-resistant bacteria arise to form a partial explanation of how this shows evidence for the theory of evolution.
Level 1 [1–2 marks]: The response has fragmented ideas about how antibiotic resistant bacteria arise that are not necessarily in the correct order of the process. Some ideas may be relevant, but there are insufficient links to form an explanation of how this shows evidence for the theory of evolution.
Level 0 [0 marks]: No relevant content.

Points that should be made:

- Mutations occur in a gene/some of the genes of the bacterial pathogen.
- These bacteria are not affected by the antibiotic (are resistant to it) and are not killed.
- This resistant strain of bacteria reproduces rapidly, to produce more bacteria that are resistant to the antibiotic.
- The population of the resistant strain rises showing survival of the fittest and the selection of resistant bacteria, which can then produce more offspring with the beneficial genes.
- The resistant strain then spreads because people are not immune to it and there is no effective treatment.

[6 marks]

Extinction

1. **a** No extinctions recorded from 1800 to 1920 [1 mark] but then extinction rapidly increases from 1920. [1 mark]

b Climate change [1 mark]; natural disasters/meteor hit the Earth/flooding/global warming [1 mark]; loss of habitat/deforestation/pollution/invasive species/hunting (by humans)/overfishing/disease. [1 mark]

2. a Too cold/very cold **or** oxygen/microbes cannot reach it; for microorganisms/microbes/bacteria/fungi/enzyme/reaction to work. [2 marks]

b No longer exist **or** no more left or died out/all died (do not credit 'died'). [1 mark]

Selective breeding

1. a Any **three** from: to increase disease resistance in food crops; to produce animals that produce more meat or milk; to promote certain features in domestic animals e.g. dogs with a gentle nature; to produce plants with large or unusual flowers or large fruits. [3 marks]

b Any **two** from: causes inbreeding so variation is lost; can cause health problems e.g. beef cows have so much muscle mass that they cannot move; it is cruel/harmful/unnatural/unethical. [2 marks]

2. Breed the low-fat milk producing cow with a bull [1 mark]; select offspring of cows that produce lowest fat milk [1 mark]; continue this process over several generations until all cows produce low-fat milk. [1 mark]

3. For: Any **two** from: increases value of offspring; health problems can be treated; out-breeding can be used to increase variation; breeders should be able to do this if they wish/high demand for that breed; breed has been around for long time/wrong to consign the breed to extinction.

Against: Any **two** from: causes inbreeding; causes a reduction in variation; if there is a harmful gene in the population it will be in more and more offspring; health problems can arise e.g. breathing problems; it is cruel/harmful/unnatural/unethical. [4 marks]

Genetic engineering

1. To produce useful substances. [1 mark]

2. Pollen from GM crops may be transported to neighbouring crops; herbicide-resistant crops may kill non-target organisms. [2 marks]

3. a Bacterial plasmid [1 mark]; virus. [1 mark]

b Anneals/seals/splices sticky ends together. [1 mark]

c Any **three** from: reversion of virus to disease-causing form; virus could be toxic to humans or insects; virus could transfer to another species; accept reference to ethical reasons qualified e.g. why change colour of leaves when there is no nutritional value? [3 marks]

4. a Removal of (human) gene; plasmid is cut/removed from bacteria; using enzymes; gene/DNA (from human cell) added to plasmid; plasmid inserted into bacterium. [3 marks]

b Any **one** named product from: vaccines/human growth hormone/insulin/clotting factor; an appropriate advantage that is less likely to be rejected; avoids use of insulin extracted from animals/can be used by vegans; can produce large quantities. [1 mark]

Classification of living organisms

1. a *Agaricus* – genus; *bernardii* – species. [1 mark]

b Any **two** from: photosynthesise; they feed autotrophically; have chlorophyll; have cell walls (containing cellulose); multicellular. [2 marks]

c Chordata; fish. [2 marks]

2. Any **two** from: unicellular; do not have nucleus/DNA in cytoplasm; circular DNA/plasmids. [2 marks]

3. a Evolutionary tree. [1 mark]

b Scorpionflies. [1 mark]

c New species are discovered/new evidence from fossils/new evidence from DNA analysis. [1 mark]

Section 7: Ecology

Habitats and ecosystems

1. Photosynthesis, shelter, pollination, interdependence. [4 marks]

2. All the living organisms that interact within the same ecosystem. [1 mark]

3. Total number of one species in an ecosystem. [1 mark]

4. Labels (from top to bottom): individuals, population, community, ecosystem. [4 marks]
5. The Sun/light source provides the energy for photosynthesis; photosynthesis makes glucose and oxygen. The glucose serves as food for animals. The oxygen is used for respiration by living organisms living in the ecosystem. [4 marks]

Food in an ecosystem

1. Strawberry plant – producer; snail – primary consumer; hedgehog – secondary consumer; fox – predator. [4 marks]
2. An organism that makes glucose by photosynthesis. [1 mark]
3. The feeding relationships (from the food to the feeder). [1 mark]
4. Snails may increase as fewer hedgehogs to eat them. Foxes may decrease as fewer hedgehogs to eat. [2 marks]

Biotic and abiotic factors

1. Pathogens, predators. [2 marks]
2. A physical condition that affects the distribution of an organism. [1 mark]
3. Any **three** from: dissolved carbon dioxide/oxygen concentration; temperature of water; light intensity; pH of water. [3 marks]
4. A tree removing a gas from the air. [1 mark]
5. **a** 1.5–2 m. [1 mark]
   b The distribution decreases. This might be because any higher than 2 metres it would be uncovered by sea water for too long. [2 marks]
6. Marks in three bands according to level of response.

Level 3 [5–6 marks]: The response gives a clear, logical description and explanation containing accurate ideas about the pattern shown in the predator–prey cycle graph.

Level 2 [3–4 marks]: The response has some key ideas about the predator–prey cycle to form a partial description and explanation of what is happening at each stage in the cycle.

Level 1 [1–2 marks]: The response has fragmented ideas about how predators and prey are dependent on each other, but they are not necessarily in the correct order of the cycle. Some ideas may be relevant, but there are insufficient links to form an explanation of what is happening at each stage.

Level 0 [0 marks]: No relevant content.

Points that should be made:
- The rabbit population has plenty of food, so they breed and increase in number rapidly.
- The increase in the rabbit population means that there is more food for the fox so the fox breeds and also increases in number.
- There are more foxes/predators so more rabbits are eaten, and the number of rabbits rapidly decreases.
- As the number of rabbits decreases, there is less for the fox to hunt and feed on, so the number of foxes also decreases.
- As the numbers of foxes decreases, the fewer rabbits are eaten so more survive to breed and so the population of rabbits increases again.
- There is a lag between the increase of rabbits and increase of foxes because the rise of the fox population is dependent on the rise in the rabbit population.
- The cycle continues.

Adapting for survival

1. Camel – large surface area to volume ratio; polar bear – thick fur and blubber; toad – long sticky tongue. [3 marks]
2. Structural adaptations are physical features of an organism. Behavioural adaptations are the things organisms do to survive. [2 marks]
3. The stings deter herbivores from eating the nettles. [1 mark]
4. Extremophiles. [1 mark]
5. Large ears: have big surface area and can be flapped to cool down. Tusks: for digging into ground to search for food or water. [2 marks]

Measuring population size and species distribution

1. Place a tape measure 20 m across the trampled area to form a transect line. Place the 1 m² quadrat against the transect line so that one corner of it touches the 0 m mark on the tape measure. Count and record the number of buttercups within the quadrat. Repeat this process at 5-metre intervals along the transect line. Repeat this process at the untrampled area. [5 marks]

2. **a** Mean: $8 \div 5 = 1.6$; mode: 0; median: 1. [3 marks]

   b Fewer buttercups due to less light/water in soil. [2 marks]

Cycling materials

1. Carbon, nitrogen, cycled, physical. [2 marks]

2. **a** Photosynthesis. [1 mark]

   b Respiration, combustion. [2 marks]

   c They return carbon dioxide back to the atmosphere when they respire. [1 mark]

   d Bacteria and fungi need oxygen to respire and produce energy when decomposing the organic material. [1 mark] Waterlogged soils do not have much oxygen so the decomposers respire less and work more slowly. [1 mark]

3. $7.4 - 5.3 = 2.1$; $2.1/5.3 \times 100 = 39.62\%$ [2 marks]

4. Every living organism on Earth depends on water to survive. Without water and the water cycle to circulate water, all living organisms would die very quickly. It is needed for chemical reactions in living organisms such as respiration and photosynthesis. [3 marks]

Effects of human activities

1. All the variety of all the different species of organisms on earth or within an ecosystem. [1 mark]

2. Biodiversity ensures the stability of ecosystems by reducing the dependence of one species on another for food, territory or mates. [1 mark]

3. There is less food available for insect pollinators such as bees. Habitats are destroyed. [2 marks]

4. To produce garden compost. [1 mark]

5. To provide land for cattle or rice fields for food. To grow biofuel crops. [2 marks]

6. **a** $100 - 65 - 3 = 32\%$. [1 mark]

   b 1. Habitats may be destroyed by landfill site. 2. Toxic chemicals from landfill can pollute land and water. [2 marks]

7. Any **four** from: burning of trees releases carbon dioxide into the atmosphere; microorganisms that breakdown remaining plant material also produce carbon dioxide as they respire; fewer trees means less photosynthesis so less carbon dioxide removed from atmosphere; less biodiversity as fewer habitats; more erosion of soil and landslides; less transpiration of water from trees can impact microclimate. [4 marks]

Global warming

1. Carbon dioxide, methane. [2 marks]

2. Any **two** from: burning fossil fuels; more rice crops; increase in cattle farming; deforestation; destruction of peatlands; more petrol cars being used; any other appropriate answer. [2 marks]

3. The graph shows that the temperature has fluctuated but there is an overall increase in global temperature. The graph clearly shows an overall trend of increasing global temperatures/ an increase of 0.7 degrees between 1880 and 2000. This is strong evidence that global warming is happening as the temperature increase is more in recent years compared with 1880. Overall, the evidence for global warming shown by the graph is quite strong. [4 marks]

4. Any **four** from: Rise in sea level due to ice caps and glaciers melting, so loss of habitat from flooding; change in distribution of species in areas where rainfall/temperature changes. Some species might increase if they favour new conditions, but others are likely to become endangered by unfavourable conditions; changes to migration patterns; loss of wetland in the African Sahel; would cause turtles and birds to decrease in numbers and possibly become extinct. [4 marks]

Maintaining biodiversity

1. By replanting hedgerows. [1 mark]

2. Any **two** from: they provide a habitat for a wide range of plants and animals, e.g. wildflowers, insects and birds; provide wildlife corridors, allowing wildlife to move freely between habitats to find food, shelter, mates; wildflowers are important sources of nectar and pollen for insect-pollinators so field margins promote

the pollination of plant species dependent on insect-pollinators. [2 marks]

3. Less waste goes to landfill so fewer habitats destroyed by landfill sites and less land, water and air pollution associated with landfill; materials are recycled so fewer trees need to be used to make new paper; less pollution associated with manufacturing paper from wood. [3 marks]

4. Planting trees does result in more carbon dioxide being absorbed from the atmosphere during photosynthesis. Carbon dioxide emissions from human activities are high so a huge number of trees would need to be planted to offset carbon dioxide emissions. [2 marks]

5. They keep damage to food chains and food webs to a minimum. They can attract tourists and therefore benefit the economy. [2 marks]

6. Any **two** points from: local people might rely on the mangrove forests for food so might not want it to be protected; local people or building companies might not know about the importance of the mangrove so might think it doesn't matter if it gets destroyed; mangroves are complicated ecosystems that require the involvement of experts for restoration, and finding experts may be difficult/expensive. [2 marks]